List of titles

Already published

Cell Differentiation	J. M. Ashworth
Biochemical Genetics	R. A. Woods
Functions of Biological Membranes	M. Davies
Cellular Development	D. Garrod
Brain Biochemistry	H. S. Bachelard
Immunochemistry	M. W. Steward
The Selectivity of Drugs	A. Albert
Biomechanics	R. McN. Alexander
Molecular Virology	T. H. Pennington, D. A. Ritchie
Hormone Action	A. Malkinson
Cellular Recognition	M. F. Greaves
Cytogenetics of Man and other Animals	A. McDermott
RNA Biosynthesis	R. H. Burdon
Protein Biosynthesis	A. E. Smith
Biological Energy Conservation	C. Jones
Control of Enzyme Activity	P. Cohen
Metabolic Regulation	R. Denton, C. I. Pogson
Plant Cytogenetics	D. M. Moore
Population Genetics	L. M. Cook
Insect Biochemistry	H. H. Rees
A Biochemical Approach to Nutrition	R. A. Freedland, S. Briggs
Enzyme Kinetics	P. C. Engel

In preparation

The Cell Cycle	S. Shall
Polysaccharide Shapes	D. A. Rees
Microbial Metabolism	H. Dalton, R. R. Eady
Bacterial Taxonomy	D. Jones, M. Goodfellow
Molecular Evolution	W. Fitch
Metal Ions in Biology	P. M. Harrison, R. Hoare
Cellular Immunology	D. Katz
Muscle	R. M. Simmons
Xenobiotics	D. V. Parke
Human Genetics	J. H. Edwards
Biochemical Systematics	J. B. Harborne
Biochemical Pharmacology	B. A. Callingham
Biological Oscillations	A. Robertson
Photobiology	K. Poff
Functional Aspects of Neurochemistry	G. Ansell, S. Spanner
Cellular Degradative Processes	R. J. Dean
Transport Phenomena in Plants	D. A. Baker
Membrane Assembly	J. Haslam

OUTLINE STUDIES IN BIOLOGY

Editor's Foreword

The student of biological science in his final years as an undergraduate and his first years as a graduate is expected to gain some familiarity with current research at the frontiers of his discipline. New research work is published in a perplexing diversity of publications and is inevitably concerned with the minutiae of the subject. The sheer number of research journals and papers also causes confusion and difficulties of assimilation. Review articles usually presuppose a background knowledge of the field and are inevitably rather restricted in scope. There is thus a need for short but authoritative introductions to those areas of modern biological research which are either not dealt with in standard introductory text-books or are not dealt with in sufficient detail to enable the student to go on from them to read scholarly reviews with profit. This series of books is designed to satisfy this need. The authors have been asked to produce a brief outline of their subject assuming that their readers will have read and remembered much of a standard introductory textbook of biology. This outline then sets out to provide by building on this basis, the conceptual framework within which modern research work is progressing and aims to give the reader an indication of the problems, both conceptual and practical, which must be overcome if progress is to be maintained. We hope that students will go on to read the more detailed reviews and articles to which reference is made with a greater insight and understanding of how they fit into the overall scheme of modern research effort and may thus be helped to choose where to make their own contribution to this effort. These books are guidebooks, not text-books. Modern research pays scant regard for the academic divisions into which biological teaching and introductory textbooks must, to a certain extent, be divided. We have thus concentrated in this series on providing guides to those areas which fall between, or which involve, several different academic disciplines. It is here that the gap between the text-book and the research paper is widest and where the need for guidance is greatest. In so doing we hope to have extended or supplemented but not supplanted main texts, and to have given students assistance in seeing how modern biological research is progressing, while at the same time providing a foundation for self help in the achievement of successful examination results.

J. M. Ashworth, Professor of Biology, University of Essex.

Enzyme Kinetics

The Steady-state Approach

Paul C. Engel

Lecturer in Biochemistry,
University of Sheffield

LONDON
CHAPMAN AND HALL

A Halsted Press Book
JOHN WILEY & SONS, NEW YORK

First published in 1977
by Chapman and Hall Ltd
11 New Fetter Lane, London EC4P 4EE

Typeset by Preface Ltd., Salisbury, Wilts. and
printed in Great Britain at the
University Printing House, Cambridge

ISBN 0 412 15070 0

Distributed in the U.S.A.
by Halsted Press, a Division
of John Wiley & Sons, Inc., New York

Library of Congress Cataloging in Publication Data

Engel, Paul C.

Enzyme kinetics.

(Outline studies in biology)
Bibliography: p.
Includes index.
1. Enzymes. 2. Chemical reaction, Rate of.
I. Title
QP601.E513   574.1'925   77-1882
ISBN 0-470-99097-X

Contents

Dedication

This book is affectionately dedicated to my father, still active at the laboratory bench at eighty-three and a continuing source of inspiration.

Acknowledgements

I am deeply indebted to Dr Keith Dalziel, F.R.S. and Professor Vincent Massey, who, between them, taught me my trade; to Dr William Ferdinand and Professor Walter Bartley, whose thoughtful suggestions greatly improved the manuscript; to Liz, Thomas and Anna for everything else.

Note to the reader

Although rapid-reaction techniques are playing an increasingly important part in the study of enzyme mechanisms, rapid-reaction kinetics has not been included in this book. The topic has been covered admirably elsewhere. It seemed best in the limited space available in the present volume to concentrate on the steady-state approach, which is of interest from the standpoints both of mechanism and metabolism. Whereas rapid-reaction kinetics remains the preserve of the specialist, every biochemist needs a basic understanding of steady-state kinetics.

1 Introduction

1.1 The book

This book makes no claim to be a comprehensive treatise, and it is not intended to compete with books in that category. It is aimed at 'non-believers', that is to say the 90% or so of biochemistry students, and indeed of practising biochemists, who place enzyme kinetics in the same category as Latin and cold showers, character-building perhaps, but otherwise to be forgotten as quickly as possible. Kinetics often appears to bristle with forbidding symbols, and many potential converts are discouraged before they even start. I believe, however, that much of enzyme kinetics can be conveyed and understood qualitatively. Algebra is needed only subsequently to provide rigorous confirmation. Approached in this way the subject is easy to grasp, and the symbols take on real meaning.

I hope also that the book will help to persuade sceptics that enzyme kinetics is of importance to most biochemists and not just to its devotees.

1.2 What are enzymes?

Biochemistry was born in the resolution of the famous conflict between Liebig and Pasteur as to the nature of fermentation. To the chemist the process was evidently chemical. To the biologist it was demonstrably biological. Both, of course, were right, although they could not see that their viewpoints were reconcilable. It was the Buchner brothers who showed that yeast, a living organism, yielded a non-living cell-free extract capable of fermenting sugars. A long line of subsequent investigators showed that this was due to the presence in the extract of many powerful catalysts, each one highly specific for one of the steps in the breakdown of glucose. Similar catalysts, or enzymes, are found in all living things.

The chemical nature of enzymes remained in dispute for a long time. Willstätter, working with peroxidase, had chosen an enzyme of such high catalytic efficiency that he believed that his active preparations were protein-free. By contrast, Sumner's crystalline urease was of such relatively modest activity that his critics attributed the catalysis to a highly active trace contaminant rather than the purified protein itself. Improvements in fractionation procedures have since allowed the purification of many hundreds of enzymes from diverse sources, leading to the realization that all enzymes are proteins.

Proteins represent both the expression of the genetic blueprint and the means of that expression. Protein synthesis requires enzymes. This apparent paradox raises fascinating questions regarding the origins of life. It is the specificity of proteins, that is to say their capacity for selective molecular recognition, that gives to life its directionality and order. Enzymes form a rather special category, because they not only recognize substances but also transform them chemically. This enables living things actively to mould and use their environment rather than merely existing as passive accretions of preformed substances. The network of enzyme-catalysed reactions that we call metabolism is subject, moreover, to delicate control minimizing waste, and the properties of key regulatory enzymes form an important part of the control apparatus.

The study of structure has led us first to a knowledge of the 'primary' linear sequence of amino acid residues that make up a protein chain and latterly, through X-ray crystallography, to a clear picture, for some proteins at least, of the three-dimensional arrangement of such linear chains. This, however, is still only a static picture, and leads us, therefore, to the next question.

1.3 What is enzyme kinetics?

Enzyme kinetics is the study of enzymes in action. The extremely high rate of enzyme-catalysed reactions greatly facilitates this study. Consider, for example, the haem-containing proteins, haemoglobin and catalase. Haemoglobin binds oxygen. It may bind and release many oxygen molecules in the course of a minute but they remain oxygen molecules, and at any instant, only one is associated with each haem centre. Catalase, being an enzyme, has a cumulative effect. Again no more than one H_2O_2 molecule will be bound per haem, but while it is bound it may react, and one therefore observes a rapid evolution of oxygen – about a million molecules per minute per enzyme molecule.

The task of enzyme kinetics is the systematic analysis of such processes, involving a study of the dependence of reaction rates on substrate concentration, pH, temperature, ionic strength and other relevant variables.

Kinetic information is obviously of descriptive value. It permits the classification and distinction of individual enzymes. Hexokinase and glucokinase, for instance, both catalyse the phosphorylation of glucose by ATP. No doubt we shall be able before long to specify the difference between these two proteins precisely in terms of their detailed structures, but the metabolic biochemists' longstanding interest in hexokinase and glucokinase stems from a recognition of their *functional* (i.e. kinetic) differences.

To quote another example, enzyme evolution may be simulated by applying selective nutritional pressures [1,2] to bacterial cultures. Mutant enzymes are quickly identified by their distinctive kinetic properties.

The intelligent use of enzymes as tools, for instance for specific metabolite assays in research, clinical diagnosis or food analysis [3], requires sound knowledge of their kinetic behaviour. Purified enzymes are also being applied increasingly in biochemical engineering [4]. Enzyme 'reactors', in which enzymes are covalently attached to solid supports can be used for syntheses without side reactions. The design of such reactors is evidently a kinetic problem.

It is also sometimes useful to subvert normal enzyme function. This can often be done selectively through the use of 'active-site-directed' irreversible inhibitors. Work on such inhibitors began in a military context with nerve gases but has since been valuable in the attempt to build a rational pharmacology. It is still probably true that most chemotherapeutic agents have been stumbled upon by chance, but many drugs are in fact enzyme inhibitors. Once their site of action is identified, they may be improved more systematically, as with the anti-depressant monoamine oxidase inhibitors [5]. Direct studies of the effectiveness of an inhibitor depend, of course, on kinetic measurements (see Chapter 3).

Beyond these purely practical considerations there are the insights that kinetics provides into an enzyme's mechanism of action (to be dealt with in subsequent chapters) and its place in the overall metabolic pattern. In establishing the quantitative significance of a metabolic pathway it is not sufficient to show that the necessary enzymes are present. Each enzyme must also be capable of handling the observed throughput of metabolites. Tempest and his colleagues [6] recently challenged the accepted view that glutamate dehydrogenase is responsible for ammonia assimilation in bacteria by showing that in several species the effective activity of this enzyme is quite inadequate for the task. This led to the discovery of a new pathway and a new enzyme, glutamate synthase, which appears also to be important in plants.

As an example of the tailoring of enzymes to a metabolic role one may consider the aldolases of muscle and liver [7] which both catalyse reactions (1a) and (1b).

$$\text{fructose 1,6-diphosphate} \rightleftharpoons \text{glyceraldehyde-3-phosphate} + \text{dihydroxyacetone phosphate} \quad (1a)$$

$$\text{fructose 1-phosphate} \rightleftharpoons \text{glyceraldehyde} + \text{dihydroxyacetone phosphate} \quad (1b)$$

Muscle aldolase is primarily concerned with the catabolism of glucose via the glycolytic pathway, which requires the splitting of fructose 1,6-diphosphate (reaction (1a) in the forward direction). Liver, however, has to make glucose, and also deals with dietary fructose. This it does by making fructose 1-phosphate, splitting it (reaction (1b)), phosphorylating the resultant glyceraldehyde, and recombining the two triose phosphates to give fructose 1,6-diphosphate (reaction (1a) in reverse). A kinetic comparison of liver and muscle aldolases shows that the liver form is much better adapted for the back reaction, and that in the forward direction it is much better than the muscle enzyme at catalysing reaction (1b).

It must also be obvious that the overall regulation and integration of metabolism can only be understood through a study of the control characteristics of the separate components. Since enzymes are the working parts of the metabolic machine, their properties must define the performance of that machine. We have accumulated a large body of information about the kinetics of individual enzymes, but are only gradually feeling our way towards adequate descriptions of integrated systems involving many enzymes.

1.4 To purify or not to purify?

It is difficult to be certain that a given chemical conversion is due to the action of a single enzyme unless the enzyme is pure. This is especially true where a reaction forms part of an extended metabolic sequence. It was originally thought, for instance, that the beta-oxidation of fatty acids might involve conversion of the saturated acid to a beta-hydroxy acid in a single step. Only by separating the various enzymes was it ultimately established that the conversion involves a dehydrogenation and a hydration catalysed by different enzymes, and that these enzymes work not on the free fatty acids but on their thioesters.

The substrates and products of an enzyme under study are usually also substrates for other enzymes, and this may seriously hinder measurements of activity in crude cell extracts. The enzymologist aims, therefore, at a 'pure' preparation as judged by various criteria – ultracentrifugation, electrophoresis, chromatography, constant specific activity, immunological assay, absence of known contaminant activities etc.

This approach is always open to the criticism that it is 'non-physiological'; the enzyme may have been irreversibly altered during purification, which sometimes involves harsh treatment with heat, organic solvents, acid or other drastic agents. *In vitro* conditions, moreover, provide, at best, a poor approximation to the *in vivo* environment. Membrane-bound enzymes such as succinate dehydrogenase present especial difficulty: the problems of solubilization and purification are such that different laboratories frequently obtain *in vitro* preparations with different properties.

This is an inherent and inescapable dilemma in biochemistry, akin to Heisenberg's Uncertainty Principle. By the very act of investigating, we disturb and perhaps distort that which we wish to investigate. We can be absolutely certain however, that if we do not investigate, we shall find out nothing. One always hopes, therefore, that the results of careful *in vitro* work may lead to a first approximation of events *in vivo*, thereby indicating the relevant definitive experiments to be done on intact systems. The striking progress in biochemistry over the past 40 years is a vindication of that hope.

It is worth adding that apparent discrepancies between *in vitro* and *in vivo* work often reflect the imperfection of our grasp of metabolism rather than the disruptive nature of our purification methods. Mammalian glutamate dehydrogenase provides an example. The purified enzyme works almost equally well with NAD(H) or NADP(H), an unusual property among nicotinamide nucleotide-linked enzymes. It was claimed, nevertheless, for some years that in the mitochondrion this enzyme uses only NADPH and not NADH. This startling discrepancy turns out to be due to differences in the availability of the two coenzymes in the mitochondrion under the conditions of experimentation, rather than to alteration of the properties of the purified enzyme [8].

1.5 Methods of measuring enzyme activity

The next problem is that of how best to measure the activity of the enzyme to be studied. One may wish ultimately to study the activity of the purified enzyme over a wide range of conditions, but in order to purify it in the first place one needs a single set of conditions for convenient, sensitive, specific and reproducible monitoring of activity. A suitable buffer must be chosen, preferably one in which the enzyme is reasonably stable. One must decide in which direction to measure the reaction, if it is reversible. It may be necessary, especially in the early stages of purification to include controls to check for interfering enzyme activities. The most important element in the design of an assay, however, is the selection of a suitable physical or chemical technique for following the appearance or disappearance of one of the reactants. An assay that allows continuous recording is preferable to one that requires repeated sampling of the reaction mixture. For this reason, assays that depend on optical properties (light absorption, fluorescence or rotation) are the most widely used. If the normal reactants have no convenient optical properties, it may be possible to substitute an artificial chromophoric substrate or one which yields a chromophoric product. This device is often used in the study of hydrolases. For instance, in the case of trypsin, an endopeptidase, neither the natural substrates nor the natural products lend themselves to simple, reliable measurements, but the enzyme also works well with low molecular weight 'model' substrates such as the nitrophenol ester of N-α-carbobenzoxylysine, which gives a large optical density change upon hydrolysis.

An alternative is to devise a 'coupled assay' in which one or more auxiliary enzymes are present in order to form an ultimate product that does have useful optical properties. Thus glutamate-oxaloacetate transaminase, which catalyses reaction (2), is conveniently assayed by including in the reaction mixture malate dehydrogenase, which catalyses reaction (3), and the missing reactant, NADH. As fast as the oxaloacetate is produced by reaction (2), it is removed by reaction (3), and one can therefore monitor the rate by following the decrease in the strong absorption or fluorescence of NADH. It is important in such an assay to ensure that the coupling reaction never becomes rate limiting [9].

$$\begin{matrix}\text{L-aspartate} \\ \text{+2-oxoglutarate}\end{matrix} \rightleftharpoons \begin{matrix}\text{oxaloacetate} \\ \text{+L-glutamate}\end{matrix} \qquad (2)$$

$$\begin{matrix}\text{oxaloacetate} \\ \text{+NADH} + \text{H}^+\end{matrix} \rightleftharpoons \begin{matrix}\text{L-malate} \\ \text{+NAD}^+\end{matrix} \qquad (3)$$

Whatever the method of assay, the time course of reaction will be curved; as the concentrations of substrates decrease, and products

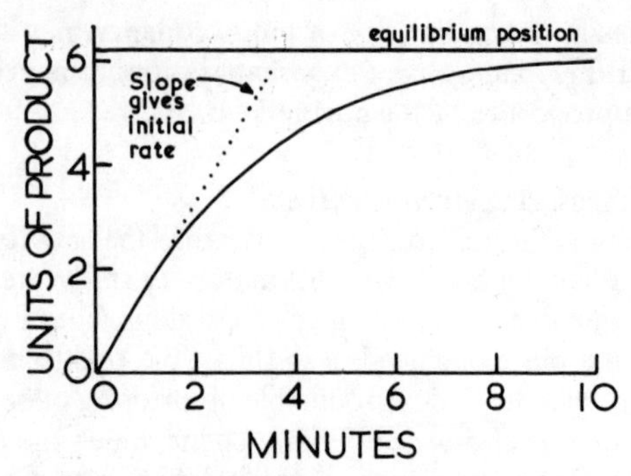

Fig. 1.1 Typical reaction time-course

accumulate, the net forward rate decreases, approaching zero at infinite time (Fig. 1.1). The rate of reaction at any given time is the slope of the tangent to this curve. Most assays are based on an estimate of the 'initial rate', the slope of the tangent at time zero. Tangents to an experimentally obtained curve, often overlaid with instrumental 'noise', are not easy to draw accurately, however, and the first few seconds of reaction are usually lost because of the time taken for mixing. Conditions are therefore chosen to minimize the curvature. This may often be achieved simply by decreasing the concentration of enzyme and increasing the sensitivity setting on the detecting instrument (Fig. 1.2). Occasionally the curvature is due to enzyme instability rather than substrate depletion, in which case the stratagem just mentioned is likely to make matters worse. One must then either add a suitable stabilizing agent or change the buffer completely.

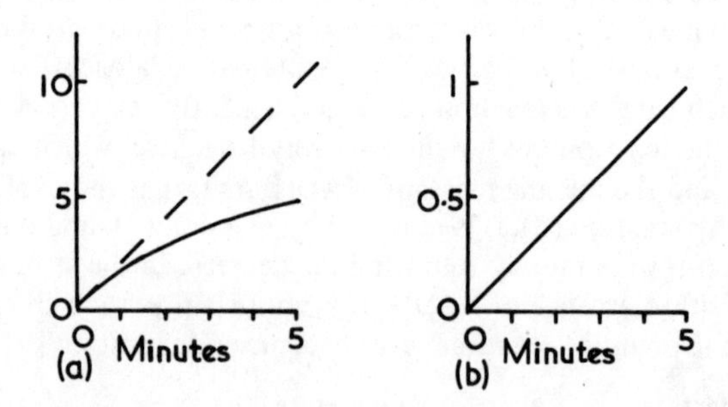

Fig. 1.2 Influence of instrumental sensitivity and enzyme concentration on the determination of initial rates. The trace in (a) is obviously too curved for accurate estimation of the initial rate. In (b) the instrumental sensitivity has been increased 10-fold so that the early linear part of the time course occupies the full vertical expanse, and only 1/10 as much enzyme is added, so that the linear portion is also stretched out along the time axis to give optimal display with a slope of about 45°.

To allow comparison between experiments, crude rates must be converted to 'specific' rates. For an impure preparation this involves dividing by the protein concentration or some other relevant indicator of the amount of material added; for a pure enzyme, if the molecular weight is known, the molar concentration of enzyme may be used instead. The conversion in either case implies the important assumption that the rate is strictly proportional to enzyme concentration. This has to be tested. If the assumption turns out to be unjustified, this may mean that the enzyme preparation contains an inhibitor or activator, or possibly that it dissociates into subunits upon dilution. Dixon and Webb ([39] p. 56–63) discuss a number of cases of apparent non-proportionality.

In the past, enzyme activity has often been expressed in arbitrary and mysterious units. The Enzyme Commission, seeking standardization, introduced the international 'Unit' of enzyme activity, that amount of enzyme which converts *one μmole of substrate per minute* under standard conditions of assay. This is now widely used. Specific activities are expressed on this basis in Units per mg. protein. More recently [10] the Commission has suggested that the old 'Unit' be replaced by the 'katal', the amount of enzyme that converts *one mole of substrate per second* (1 Unit = 16.67 nkat). Specific activities should now be given in katals per kilogram.

Finally it must be emphasized that kinetic properties of different enzyme preparations can only be validly compared if they are determined under identical conditions – unless, of course, the variation with conditions is the object of the comparison. It is sometimes claimed that two preparations contain kinetically distinct 'isoenzymes' (different proteins catalysing the same reaction), when in fact they have been studied at different temperatures, pH values etc. Such comparisons are meaningless.

For the rest of this book we shall be concerned with the variation of enzyme-catalysed reaction rates, not as a matter of technical detail, but as a subject in its own right. We shall also assume from now on that the enzymes under discussion have been satisfactorily purified and are free of interfering activities.

2 One-substrate kinetics

2.1 Saturation kinetics

If we consider a hypothetical reaction S ⟶ P obeying first-order kinetics, the rate is given by:–

$$-\frac{d[S]}{dt} = \frac{d[P]}{dt} = k[S]$$

i.e. the chance of a molecule of P being formed at any given instant is directly proportional to the concentration of S. Similarly, for a second-order reaction S + T ⟶ P + Q, the rate would be proportional to the concentration of each of the reactants S and T:–

$$-\frac{d[S]}{dt} = -\frac{d[T]}{dt} = k[S]\,[T].$$

In either case a plot of *initial* rate against varied concentration of S would be a straight line, of slope k in the first case and $k[T]$ in the second (Fig. 2.1). (Chemical kineticists tend, in fact, to use the full time course of reaction rather than the initial rate.)

If these reactions were enzyme-catalysed, the dependence of the initial rate on substrate concentration would be different. The rate would probably be proportional to [S] for low values of [S], but with higher values of [S] the rate would asymptotically approach a maximum. This is called saturation. The plot of initial rate against [S] is hyperbolic (Fig. 2.2).

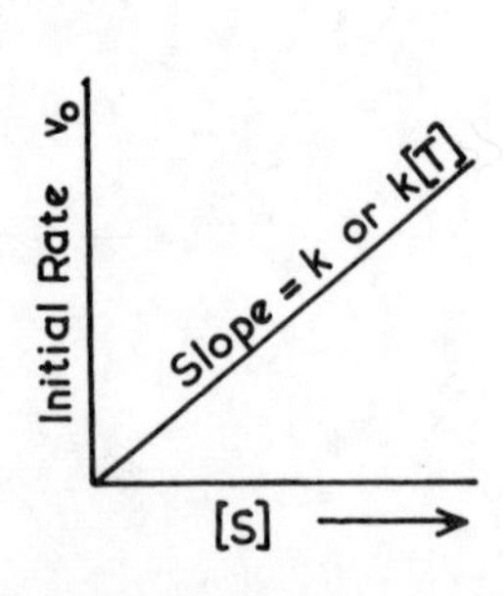

Fig. 2.1 Dependence of initial rate on reactant concentration for a simple first- or second-order chemical reaction.

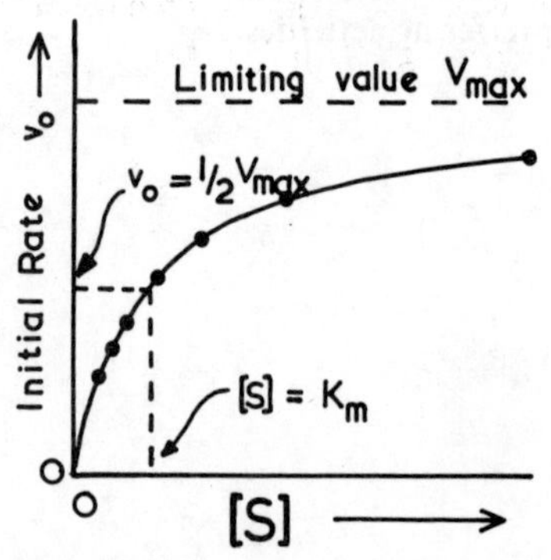

Fig. 2.2 Dependence of initial rate on substrate concentration for a typical enzyme-catalysed reaction.

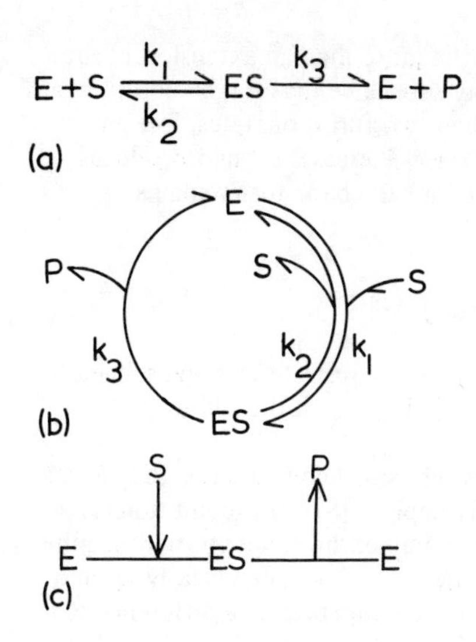

Fig. 2.3 Three different schematic representations of a one-substrate mechanism.

The major credit for realizing the the significance of this pattern goes to A. J. Brown who proposed in 1902 that invertase forms a stoicheiometric complex with its substrate [11]. It is only this complex, he suggested, that breaks down to the products, simultaneously regenerating free enzyme (Fig. 2.3). The nearest analogy to this pattern in classical chemistry is that of surface catalysis.

Brown's kinetic inference of the existence of a one-to-one enzyme-substrate complex was made long before the chemical nature of enzymes was known, 40 years before the spectrophotometric detection of such complexes, and 60 years before their 'visualization' through X-ray crystallography. In the intervening period this one concept has provided the foundation for our understanding of biochemical specificity. This despite the fact that, like all inference from kinetics, it was tentative, consistent with the facts rather than logically inescapable.

The reaction scheme in Fig. 2.3a may also be depicted as in Fig. 2.3b to emphasize the cyclic nature of catalysis, or in the notation developed by Cleland (Fig. 2.3c), which highlights the enzyme-containing complexes rather than the chemical reactants.

Now in this mechanism the initial rate of formation of P is proportional to the concentration of ES, the enzyme-substrate complex. At low concentrations of S the small amount of ES formed is proportional to [S], and so the overall rate is also proportional to [S]. As [S] is raised, however, the balance between E and ES shifts until virtually all the enzyme is present as ES, i.e. the enzyme is 'saturated' with S. In this state the enzyme is working flat out. As fast as ES breaks down to release E and P, or E and S, more S jumps on to re-form the enzyme-substrate complex. Since the enzyme is present at a fixed concentration, e, the rate cannot exceed k_3e (Fig. 2.2).

The whole of enzyme kinetics is little more than an extension of the simple idea introduced above, that any substance that can be bound to an enzyme distributes the enzyme between two forms or states, free and complexed. The number of different states increases if one introduces inhibitors, or more than one substrate, but the basic idea remains unaltered.

2.2 The Michaelis-Menten equation.

A simple mathematical treatment of the one-substrate reaction scheme was given by Michaelis and Menten [12]. It rests on the following assumptions:–

(i) the concentration of enzyme is very small compared to [S], so that formation of ES does not significantly deplete [S]. This condition is met in most catalytic experiments. If, for example, the concentration of substrate is 10^{-3} M and that of the enzyme is 10^{-9} M, then clearly, even if all the enzyme is present as ES, it makes an imperceptible difference to [S].

(ii) the concentration of P is effectively zero. This is the 'initial-rate' assumption, and implies not only that P is absent at the outset, but also that the amount of P formed in the time required for a rate measurement is too small to give rise to a significant reverse reaction;

(iii) although the product-releasing step is fast, it is nevertheless so much slower than the alternative reaction in which S is released from ES, that E and ES may be considered to be effectively at equilibrium. This third, somewhat arbitrary assumption distinguishes the Michaelis-Menten treatment from that later adopted by Briggs and Haldane.

Setting up the equilibrium expression we have:–

$$k_1 [E] [S] = k_2 [ES]$$

$$\therefore [E] = [ES] \frac{k_2}{k_1 [S]} \tag{2.1}$$

But we also have another equation relating the two unknown variables, [E] and [ES]. It is the so-called 'enzyme conservation equation', which states in essence that we know how much enzyme was put in, although we may not know precisely where it has gone. Thus:

$$[E] + [ES] = e \tag{2.2}$$

where e is the total enzyme concentration. We can now substitute for [E] from equation (2.1):

$$\therefore e = [ES] \frac{k_2}{k_1 [S]} + [ES] = [ES] \left(1 + \frac{k_2}{k_1 [S]}\right). \tag{2.3}$$

Now the overall rate, v, is equal to $k_3 [ES]$, since the reverse rate may be ignored.

$$\therefore v = \frac{k_3 e}{1 + \dfrac{k_2}{k_1 [S]}} = \frac{k_3 e [S]}{[S] + k_2/k_1} \tag{2.4}$$

This gives the rate as a function of e and [S], both known, and the constants k_1, k_2 and k_3. Equation (2.4) can be trimmed down a bit further: since k_1 and k_2 are constant, their ratio is constant and can be replaced by K_m, the Michaelis constant. We also know something about $k_3 e$: it is the maximum rate, V_{max}. Thus we can rewrite the Michaelis-Menten equation in its conventional form:

$$v = \frac{V_{max} [S]}{[S] + K_m} \tag{2.5}$$

If the derivation is checked, one can see that the left-hand side of the denominator in equation (2.5), [S], represents, or is proportional to, the contribution of ES to the total, whilst the right-hand side represents the contribution of E, the free enzyme. Accordingly, as [S] becomes very large compared to K_m, and [ES] becomes very large compared to [E], v approaches $V_{max}[S]/[S]$, i.e. V_{max}. On the other hand, when [S] is very small compared to K_m, and the enzyme is almost all present as E, $v = V_{max}[S]/K_m$, i.e. the rate is proportional to [S]. This coincides with our earlier, qualitative prediction. These are the two extremes. What happens when $[S] = K_m$? Equation (2.5) then becomes

$$v = \frac{V_{max} [S]}{[S] + [S]} = \frac{V_{max}}{2}$$

Hence K_m *is that substrate concentration which gives half the maximum rate* (Fig. 2.2). This is the usual operational definition of K_m.

2.3 Linear plots

In order to use equation (2.5) one needs to know the values of V_{max} and K_m. To obtain these values one measures the initial rate with several different concentrations of S. It is not easy, however, to draw a hyperbola through a set of experimental points, especially if the measurements extend only up to, say, 80% V_{max} – perhaps because the substrate is rather insoluble, or expensive! Without an accurate extrapolated value of V_{max} one cannot obtain $\frac{1}{2}V_{max}$ in order to estimate K_m.

This practical problem has led kineticists to re-arrange equation (2.5) in linear forms. Of the various forms introduced by Woolf (13), the 'Lineweaver-Burk plot' [14] has become the most popular. Taking reciprocals of both sides of equation. (2.5),

$$\frac{1}{v} = \frac{K_m + [S]}{V_{max}[S]} = \frac{K_m}{V_{max}[S]} + \frac{[S]}{V_{max}[S]}$$
$$= \frac{K_m}{V_{max}[S]} + \frac{1}{V_{max}}. \tag{2.6}$$

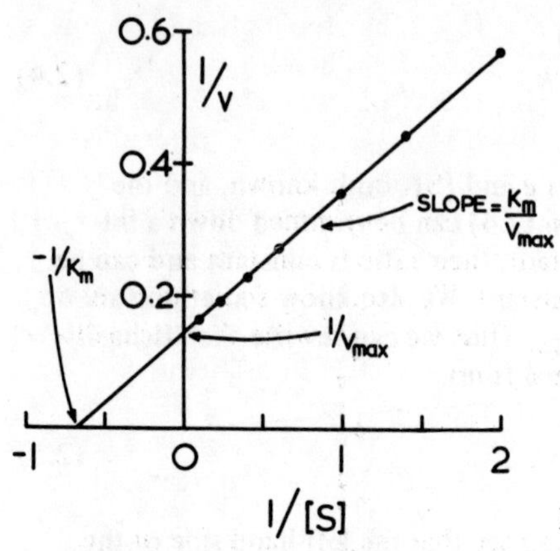

Fig. 2.4 A Lineweaver–Burk plot.

This is of the form $y = mx + c$ (the general equation for a straight line) where y is $1/v$ and x is $1/[\mathrm{S}]$. Thus the Lineweaver-Burk plot of $1/v$ against $1/[\mathrm{S}]$ is linear, with a slope of K_m/V_{max} (Fig. 2.4). The intercept on the ordinate is $1/V_{max}$, as may be seen by substituting $1/[\mathrm{S}] = 0$, or $[\mathrm{S}] = \infty$, in equation (2.6). This is to be expected: when $[\mathrm{S}] = \infty$, $v = V_{max}$, and so, when $1/[\mathrm{S}] = 0$, $1/v = 1/V_{max}$.

Substituting $1/v = 0$ in equation (2.6), we find that the negative intercept on the abscissa gives the reciprocal of K_m (Fig. 2.4).

This linearization means that, within reason, experimental data may be extrapolated to give the kinetic constants. Two situations in which this is not possible are shown in Fig. 2.5. From line A, an accurate estimate of V_{max} is readily obtainable, but there is so little variation in the rate that K_m certainly cannot be accurately estimated from the slope or the abscissa intercept. This is because the values of [S] are all near-saturating. Line B shows the opposite situation, plenty of low values of [S], allowing

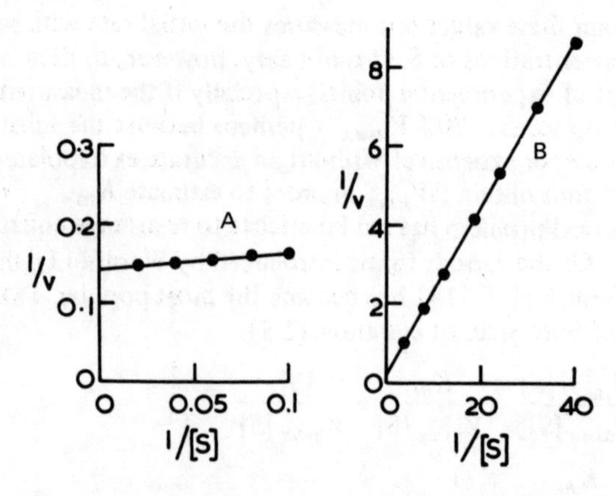

Fig. 2.5 Unsatisfactory Lineweaver–Burk plots.

accurate estimation of the slope, K_m/V_{max}, but too long an extrapolation for V_{max}. Thus we cannot obtain K_m from line B either. It is important to be able to estimate both the slope and the ordinate intercept of a Lineweaver–Burk plot with reasonable precision. The range of substrate concentrations should therefore span the K_m, with the highest values, ideally, approaching saturation, and the lowest values giving a rate of, say, $0.25 \times V_{max}$ or less (Fig. 2.4).

The only conceptual barrier to the use of Lineweaver–Burk plots is the difficulty of adjusting to reciprocals: the 'lowest' points represent the highest rates and values of [S]; the points furthest from the origin represent the lowest rates and values of [S].

Two other linearizations of equation (2.5) are sometimes used. The relevant equations are readily arrived at by multiplying equation 2.6 by vV_{max} in one case and by [S] in the other. The first variant leads to a plot of $v/[S]$ against v (Eadie–Hofstee plot), whereas the second leads to a plot of $[S]/v$ against [S] (Hanes plot). As an exercise it is worth working out what these plots would look like and how K_m and V_{max} would be obtained. Proponents of these plots point out that in each case one of the two variables of interest, v and [S], appears as such, and not as a reciprocal. On the other hand, the Lineweaver–Burk plot at least keeps the independent variables separate, whereas the other plots involve compound variables, $[S]/v$ and $v/[S]$, which are surely less easy to comprehend than $1/v$ or $1/[S]$.

The principal argument for the alternative plots lies in another criticism [15, 16] of the Lineweaver–Burk plot, namely that it biases the interpretation of experimental results, giving undue weight to inaccurate measurements made at low substrate concentrations, and insufficient weight to the more accurate higher rates. There are several answers to this argument:

(i) the high concentration points are not always the most accurate, especially if the rate measurement is based on the disappearance of substrate rather than the appearance of product; any deviations from linearity are also likely to occur at high substrate concentrations;

(ii) in plotting 'by hand' any experienced kineticist bears in mind the properties both of the plot and of the experimental system in weighting the points and deciding where to draw the line;

(iii) if the data are handled by a computer – e.g. with the SEQUEN programme of Cleland [17] – they are fitted to the v against [S] hyperbola, and the Lineweaver–Burk plot is then merely used as a convenient and familiar way of displaying the results;

(iv) one should aim in any case to obtain experimental data good enough to make the whole issue of weighting irrelevant. Graphical sleight of hand can never turn bad data into good data.

The choice of plotting method is, in short, a matter of personal preference, but the Lineweaver–Burk plot is still by far the most widely used, and it is important to be thoroughly familiar with it.

The three plots discussed above do not exhaust the possibilities. Some interesting new methods of plotting have been recently introduced [18, 19]. Whether they will gain favour remains to be seen. In the method of Eisenthal and Cornish–Bowden [18] each measurement is represented by a line rather than a point. This makes for a less compact presentation of the data. The statistical merits of the various methods are discussed in detail by Markus *et al.* [19].

2.4 The steady-state assumption

The assumption [12] of equilibrium between E and ES (p. 16) is unnecessarily restrictive. Briggs and Haldane showed in 1926 [21] that an equation formally similar to equation (2.5) may be derived without this assumption. They assumed instead a *steady state*, i.e. that the concentrations of E and ES remained effectively constant over the period of the rate measurement. According to the equilibrium assumption, the rate of formation of ES from E and S equals its rate of dissociation *to E and S*. The steady state assumption requires only that the rate of formation of ES should equal its rate of breakdown *in any direction*, including product formation, which need not be slow relative to the back-dissociation to E and S. A useful physical analogy is that of a large jug pouring water at a steady rate into a funnel. At first the rate of inflow greatly exceeds the rate of outflow, but this makes the water level in the funnel (cf. [ES]) rise. The rate of outflow (product formation) therefore increases until it exactly matches the inflow. Thereafter the level of water in the funnel and the rate of flow remain constant until the jug is emptied. The analogy is not perfect. The rate of formation of ES decreases as the steady state is established, because ES is formed at the expense of E. The jug's pouring rate, by contrast, is independent of the height of the water in the funnel. There is also no counterpart to the dissociation of ES to E and S: the water does not jump back from the funnel into the jug. This perhaps serves to emphasize the nature of a steady state as opposed to an equilibrium.

The steady state is still an assumption, albeit less restrictive than the equilibrium assumption, and its validity has been the subject of recurrent debate. The steady state may, however, sometimes be *directly* demonstrated, provided that the enzyme-substrate complexes differ appreciably from free enzyme in physical properties that can be conveniently measured on a millisecond timescale. After very rapid mixing of a concentrated solution of enzyme with its substrate(s), e.g. in a 'stopped-flow' apparatus, the build-up and decay of enzyme-substrate complexes is directly observed and recorded. The duration of the steady state, for a fixed enzyme concentration, is a function of the substrate concentration and the turnover capacity of the enzyme. To return to our analogy, if there is not much water in the jug it may all flow through before the level in the funnel has a chance to reach its steady state. Likewise if the neck of the funnel is very wide. It seems likely, therefore, that the steady state assumption is justified under the conditions of most experiments in which the enzyme concentration is low (in molar terms) and the substrate concentration is relatively high. Empirically, the assumption has been justified by its success in explaining and predicting kinetic patterns for many enzymes. As we shall see, it is also often possible to make independent checks on the validity of kinetic deductions.

Turning now to the mathematical consequences of the steady-state treatment, we have from Fig. 2.3:

$$\text{rate of formation of ES} = k_1\,[\mathrm{E}]\,[\mathrm{S}]$$

$$\begin{aligned}\text{rate of removal of ES} &= k_2\,[\mathrm{ES}] + k_3\,[\mathrm{ES}]\\ &= (k_2 + k_3)[\mathrm{ES}].\end{aligned}$$

Applying the steady state assumption, therefore,

$$k_1\,[\mathrm{E}]\,[\mathrm{S}] = (k_2 + k_3)[\mathrm{ES}]$$

$$\therefore\ [\mathrm{E}] = [\mathrm{ES}]\,\frac{(k_2 + k_3)}{k_1\,[\mathrm{S}]}. \tag{2.7}$$

The rest of the derivation is identical with that for the Michaelis–Menten treatment above, and, by substituting in equation (2.2) from equation (2.7) instead of equation (2.1), we obtain:

$$v = \frac{k_3 e}{1 + \dfrac{k_2 + k_3}{k_1\,[\mathrm{S}]}} = \frac{k_3 e[\mathrm{S}]}{[\mathrm{S}] + \dfrac{k_2 + k_3}{k_1}}. \tag{2.8}$$

Since k_1, k_2 and k_3 are all constants, this equation once again may be rewritten as equation (2.5), but the value of K_m, the Michaelis constant, is no longer the same. Whereas the Michaelis–Menten treatment leads to an equation in which $K_m = k_2/k_1$, in the Briggs–Haldane treatment $K_m = (k_2 + k_3)/k_1$.

Thus effectively the difference between the two treatments lies in the relative values of k_2 and k_3. These two constants determine the fate of ES. Either it can return whence it came, releasing S at a rate determined by k_2, or it can proceed to form product at a rate determined by k_3. If k_3 is very much smaller than k_2, so that the chances of reaction to yield P are small for any given molecule of ES compared to the chances of dissociation back to S, then $k_2 + k_3$ will approximate to k_2, E and ES will be virtually at equilibrium, and the product-forming step will be, as it were, a small leak out of that equilibrium system. Thus, if $k_2 \gg k_3$, the Briggs–Haldane equation reduces to the Michaelis–Menten equation. Equally plausible, however, is the alternative extreme assumption [22] that $k_3 \gg k_2$, so that K_m becomes k_3/k_1. This exposes the dubious nature of the frequent assertion that K_m reflects an enzyme's affinity for its substrate, a low K_m representing high affinity and vice versa. This is valid when the Michaelis–Menten assumption holds and $K_m = k_2/k_1$, since k_2/k_1 is the dissociation constant for ES. Clearly, however, if $k_3 \gg k_2$, K_m is bound to be much larger than the dissociation constant.

The ideal procedure would be to *test* the Michaelis–Menten assumption by measuring the K_m and the dissociation constant for S separately and comparing them. Unfortunately, for a one-substrate enzyme, this cannot be done, since one cannot mix E and S without a reaction occurring. With a two-substrate enzyme this problem does not arise, as we shall see in Chapter 5.

In summary, therefore, in the absence of other evidence, K_m should be regarded simply as an empirical constant equal to the substrate concentration that gives $\frac{1}{2}V_{max}$ under defined experimental conditions. As such it gives a useful indication of the range over which the enzyme is responsive to changes in [S]. Thus from equation (2.5) one can readily calculate that, when [S] is four times K_m, v is 80% of V_{max}, so that further increases in [S] bring only a small increase in rate.

2.5 The reverse reaction

All chemical reactions, and therefore all biochemical reactions, are in theory reversible. If the position of equilibrium lies very far over to one side, it may only be experimentally possible to demonstrate net conversion in one direction. Often, however, conditions may be chosen to favour the reverse reaction e.g. the alcohol dehydrogenase reaction (reaction 4) may be made 'quantitative' in either direction by altering the pH, since a proton is involved.

$$C_2H_5OH + NAD^+ \rightleftharpoons CH_3CHO + NADH + H^+ \qquad (4)$$

Certainly many enzyme-catalysed reactions are freely reversible, and indeed many enzymes function in both directions physiologically. This being so, we should modify our one-substrate mechanism as below

$$E + S \underset{k_2}{\overset{k_1}{\rightleftharpoons}} ES/EP \underset{k_4}{\overset{k_3}{\rightleftharpoons}} E + P$$

by including a step leading from P to the enzyme-substrate complex, and recognising that, if we have only one such complex, it is as much an EP complex as an ES complex.

In the presence of enzyme, equilibrium is established in due course. It

Table 2.1 Symmetry-related rate constants for forward and reverse one-substrate reactions

Rate constants involved in S → P	*Equivalent rate constants involved in* P → S
k_1	k_4
k_2	k_3
k_3	k_2

may be approached from either S or P. If the starting mixture contains P but no S, we may measure the initial rate of formation of S, just as we previously measured rates of formation of P from S. Now we can write down the initial rate equation for the formation of S from P *by inspection*: the reaction scheme is entirely symmetrical. For each constant, therefore, we may write down a 'mirror-image' constant as shown in Table 2.1. This Table may then be used as a 'dictionary' to obtain the reverse rate equation, given that we already know the forward rate equation.

Thus, if we use the Briggs-Haldane approach* and 'translate' equation (2.8), we obtain

$$v_r = \frac{k_2 e[\mathrm{P}]}{[\mathrm{P}] + \dfrac{k_3 + k_2}{k_4}} \tag{2.9}$$

where v_r is the initial rate in the reverse direction. From equation (2.9), $V_{\max(r)} = k_2 e$ and

$$K_{m(r)} = \frac{k_2 + k_3}{k_4}.$$

We may now derive a useful relationship between the kinetic constants and the overall equilibrium constant, which is a thermodynamic constant characteristic of the reaction and entirely independent of the properties of the enzyme. At equilibrium there is no net reaction.

$$\therefore k_2\ [\mathrm{ES}] = k_1\ [\mathrm{E}]\ [\mathrm{S}] \qquad \therefore \frac{[\mathrm{ES}]}{[\mathrm{E}]} = \frac{k_1}{k_2}\ [\mathrm{S}]$$

Also

$$k_3\ [\mathrm{ES}] = k_4\ [\mathrm{E}]\ [\mathrm{P}] \qquad \therefore \frac{[\mathrm{ES}]}{[\mathrm{E}]} = \frac{k_4}{k_3}\ [\mathrm{P}].$$

From these two equations,

$$\frac{k_1}{k_2}\ [\mathrm{S}] = \frac{k_4}{k_3}\ [\mathrm{P}]$$

$$\therefore \quad \frac{[\mathrm{P}]}{[\mathrm{S}]} = \frac{k_1 k_3}{k_2 k_4} = K_{eq}. \tag{2.10}$$

Now, if we examine the kinetic constants for the forward and reverse enzyme-catalysed reactions, we discover that

$$\frac{V_{\max(f)}}{V_{\max(r)}} = \frac{k_3}{k_2} \tag{2.11}$$

*We are obliged to use the Briggs–Haldane approach here; the Michaelis–Menten assumption for the reverse reaction would require that $k_3 \gg k_2$, i.e. the precise opposite of the corresponding assumption for the forward reaction.

and

$$\frac{K_{m(r)}}{K_{m(f)}} = \frac{k_1}{k_4}. \tag{2.12}$$

Multiplying equations (2.11) and (2.12) together, we obtain

$$\frac{V_{max(f)}\,K_{m(r)}}{V_{max(r)}\,K_{m(f)}} = \frac{k_1 k_3}{k_2 k_4} = K_{eq}. \tag{2.13}$$

Relationships of this type, between the kinetic parameters for a reversible enzymatic reaction and the overall equilibrium constant, are known as Haldane relationships.

2.6 Integrated rate equation

When, in estimating an initial rate, one discards the rest of the reaction time-course, one is discarding a lot of information. If the curvature of the time-course reflects only the progress of the reaction, and not enzyme instability, then it should be possible to generate the curve mathematically, given the kinetic constants and the relevant concentrations. The reverse procedure is also possible. Such analysis requires the use of integrated forms of the rate equations, i.e. forms that predict the total extent of reaction at any given time after mixing. Detailed treatment (see e.g. [23]) is beyond the scope of this book, but it is worth noting that analysis of this kind can give good agreement with the results of steady-state initial-rate analysis e.g. [24,25] .

2.7 Metabolic significance of kinetic constants

The Michaelis-Menten terminology provides a simple way of describing and comparing the rate behaviour of enzymes. As we shall see in Chapter 5, caution is necessary when kinetic constants determined *in vitro* are applied in a metabolic context. Nevertheless, the K_m and V_{max} for an enzyme do give some indication of its capabilities. This may be seen by returning to some of the examples mentioned in Chapter 1. We can now be more specific, for instance, about the kinetic distinction between mammalian hexokinases and glucokinase. Hexokinases have a K_m for glucose of 10^{-5} M or less. Glucokinase, which is confined to liver, has a K_m of 10–20 mM, i.e. at least 1000-fold higher, enabling it to respond to the changing glucose level in the portal blood.

The conclusion, also quoted in Chapter 1, that glutamate dehydrogenase cannot cope with ammonia uptake in certain bacterial species is based on two lines of kinetic argument: firstly, the enzyme's K_m for ammonia is much higher than the levels of ammonia likely to be encountered by the bacteria, so that the rate of uptake through this reaction would be insensitive to changes in ammonia concentration; secondly, given the amount of glutamate dehydrogenase present in the cells, the V_{max} is much too low to account for observed rates of ammonia assimilation even if the enzyme could somehow be saturated.

The mathematical description of the kinetics of individual enzymes should allow us to move towards a mathematical description of metabolic pathways. Among the attempts that have already been made in this direction is the study by Rapoport and his colleagues of glycolysis in erythrocytes [26,27].

3 Inhibitors, activators and inactivators

3.1 Introduction

The rate of an enzyme-catalysed reaction may sometimes be altered in a specific manner by compounds other than the substrate(s). Activators increase the rate; inhibitors and inactivators decrease it. The study of such agents is of practical importance for several reasons:

(i) inhibition and activation of enzymes by key metabolites provides the normal means of metabolic 'fine control' superimposed on the 'coarse' control achieved by regulation of the synthesis and breakdown of active enzymes;

(ii) external interference with metabolism, whether by drugs, pesticides etc. or by undesirable toxic agents, often depends on the inhibition of enzymes;

(iii) inhibitors, and especially inactivators, provide a powerful tool for studying the chemical mechanisms of enzyme action.

3.2 Reversible and irreversible inhibition

Enzyme activity can be curtailed by various non-specific agents (acid or alkali, urea, detergents, proteases etc.) which disrupt protein structure. We are concerned here with more selective agents which interact with a protein at a small number of loci without markedly disrupting the three-dimensional structure. The two most important considerations in their classification are specificity and reversibility. Most of this chapter is about inhibitors that are very specific and fully reversible, but we shall first consider some other patterns that arise.

CASE 1

Iodoacetate and glyceraldehyde 3-phosphate dehydrogenase. This is a clearcut case of irreversible inactivation [28] .An essential thiol group in the enzyme is alkylated:

$$ESH + ICH_2COOH \longrightarrow ESCH_2COOH + HI. \qquad \text{... (5)}$$

The inactivation is progressive and cannot be reversed by dialysis. The addition of competing thiol compounds may slow down the inactivation, but cannot reverse it. The reaction is very specific in one sense: out of all the enzyme's amino acid residues (about 330, including 4 cysteines), only cysteine 149 reacts rapidly [28] . On the other hand, iodoacetate is not a specific reagent for this enzyme, nor for cysteine. Pancreatic ribonuclease, for instance, is inactivated by reaction of iodoacetate with its essential histidine residues.

CASE 2
5,5′-dithiobis-(2-nitrobenzoic acid). This reagent, usually referred to as DTNB or Ellman's reagent, reacts with thiol groups in many enzymes. In the reaction, one thionitrobenzoate moiety is released as the free, yellow anion, while the other becomes disulphide linked to the enzyme:

$$ES^- + O_2N\text{-}C_6H_3(COO^-)\text{-}S\text{-}S\text{-}C_6H_3(COO^-)\text{-}NO_2$$

$$\rightleftharpoons ES\text{-}S\text{-}C_6H_3(COO^-)\text{-}NO_2 + O_2N\text{-}C_6H_3(COO^-)\text{-}S^- \qquad \ldots(6)$$

This thiol-disulphide interchange is freely reversible, but once the free thionitrobenzoate is removed, the modification is stable. If an essential thiol is modified, the enzyme loses activity. Unless secondary changes ensue*, activity may be restored by adding an excess of a thiol compound, e.g. cysteine, to break the disulphide linkage. And yet this is not what is normally meant by reversible inhibition.

CASE 3
Pyridoxal phosphate and glutamate dehydrogenase. Although pyridoxal phosphate is an essential prosthetic group (see Chapter 5) for some enzymes, its reactive aldehyde group also makes it inhibitory for many others. Most frequently it reacts with the $-NH_2$ groups of lysine residues:

$$-\underset{H}{C}{=}O + H_2N- \rightleftharpoons -\overset{H}{\underset{OH}{C}}-\underset{H}{N}- \rightleftharpoons -\underset{H}{C}{=}N- + H_2O. \qquad \ldots(7)$$

Schiff base or imine

In bovine glutamate dehydrogenase, out of 500 residues, just one lysine reacts in this way [29]. This reflects in part the unusual reactivity of this residue, and in part the selectivity conferred by the pyridinium ring and the charged phosphate group of the reagent. The reaction, though covalent, can be reversed by amino compounds or even by dialysis. Even this however, is not reversible inhibition as normally understood.

*Cysteine 149 of glyceraldehyde 3-phosphate dehydrogenase is a case in point: following the initial, reversible inactivation by reaction of DTNB with this residue, an intramolecular reaction leads to an *irreversible* change [31].

CASE 4
Malonate and succinate dehydrogenase. Reversible inhibition is well exemplified by the classic case of malonate, which, as an inhibitor of succinate dehydrogenase, played a crucial role in the dicovery of the Krebs citric acid cycle [30]. The inhibition involves no covalent bonds, it is instantaneous rather than progressive, and it can be reversed either by dialysis, or, more immediately, by dilution. It is also highly specific for reasons explained below.

Having defined reversible inhibition, we shall now sub-divide our classification by considering different patterns of reversible inhibition. In the simple one-substrate case, a hypothetical inhibitor might be able to combine with E but not with ES, or with ES but not E, or with both, but perhaps with different affinities. If the inhibitor, I, can combine with ES to give a complex EIS, then one must also consider whether EIS can evolve a product, and, if so, how fast.

3.3 Competitive inhibition

Let us first examine the case in which free enzyme, E, can combine either with S, to give the productive complex ES, or with I, to give EI, but not with both. The commonest reason for this pattern is that I is a substrate analogue, and sits precisely where the substrate should be, in the enzyme's active site (Fig. 3.1). Malonate is an example of this type of inhibitor for succinate dehydrogenase.

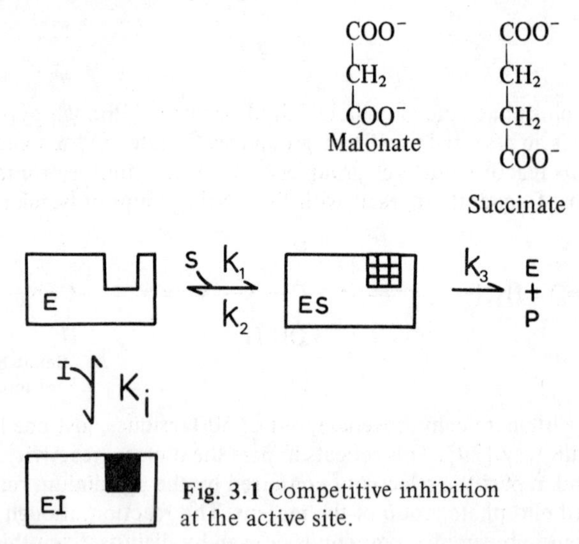

Fig. 3.1 Competitive inhibition at the active site.

In theory, however, a compound dissimilar to the substrate might bind to the enzyme elsewhere, and nevertheless alter its shape or charge distribution in such a way that the substrate could no longer bind at the active site (Fig. 3.2).

In either case S and I are mutually exclusive. With a given substrate

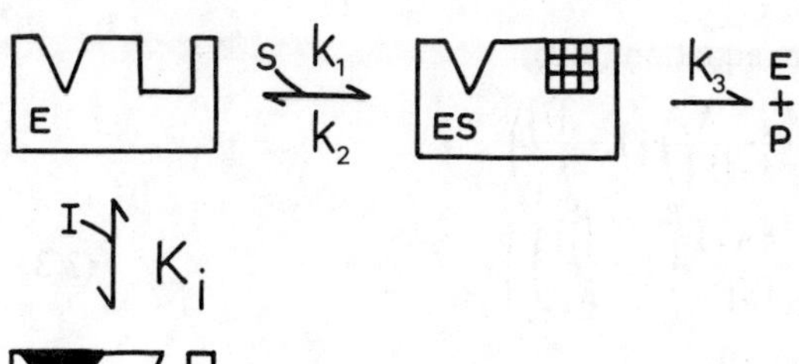

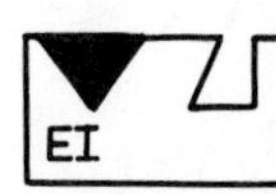

Fig. 3.2 Competitive inhibition through conformational change.

concentration, enzyme is distributed between E and ES as described in Chapter 2. Adding I converts some of the E to EI. This disturbs the balance between E and ES which readjusts in favour of E. The net effect when balance is re-established is to reduce the steady-state level of the productive complex ES, and therefore decrease the rate. Increasing the inhibitor concentration, [I], ultimately pulls all the enzyme over into the EI form, shutting down catalysis completely.

Conversely, if we increase [S], this pulls E back towards ES, EI has to release inhibitor to maintain equilibrium with E, and, with a high enough level of S, even in the presence of inhibitor essentially all the enzyme will be present as ES. The maximum rate is therefore unaffected. Precisely because I can displace S, S can also displace I. In the presence of I however, a higher level of S is required in order to approach V_{max}, and accordingly K_m is raised. Such inhibition, for obvious reasons, is known as *competitive inhibition*.

Having considered competitive inhibition in qualitative terms, let us see how the algebraic description matches our predictions. Since E is in a steady state and there is no path out of EI except back to E (Figs. 3.1 and 3.2), it is reasonable to assume that E and EI will be in equilibrium. Thus we may write:

$$\frac{[\mathrm{E}]\,[\mathrm{I}]}{[\mathrm{EI}]} = K_i \quad \text{or} \quad [\mathrm{EI}] = [\mathrm{E}]\,\frac{[\mathrm{I}]}{K_i} \tag{3.1}$$

where K_i is the dissociation constant for the enzyme-inhibitor complex. As in the absence of inhibitor, the steady-state equation for ES gives:

$$[\mathrm{ES}]\,(k_2 + k_3) = [\mathrm{E}]\,k_1\,[\mathrm{S}] \tag{3.2}$$

However, whereas previously we had

$$e = [\mathrm{ES}] + [\mathrm{E}]$$

we now have for the enzyme conservation equation

$$e = [\mathrm{ES}] + [\mathrm{E}] + [\mathrm{EI}].$$

Substituting for [EI] from equation (3.1),

$$e = [\mathrm{ES}] + [\mathrm{E}]\left(1 + \frac{[\mathrm{I}]}{K_i}\right)$$

Substituting for [E] from equation (3.2),

$$e = [\mathrm{ES}] + [\mathrm{ES}]\left(\frac{k_2 + k_3}{k_1[\mathrm{S}]}\right)\left(1 + \frac{[\mathrm{I}]}{K_i}\right)$$

$$= [\mathrm{ES}]\left[1 + \frac{(k_2 + k_3)}{k_1\,[\mathrm{S}]}\left(1 + \frac{[\mathrm{I}]}{K_i}\right)\right]. \tag{3.3}$$

As in the absence of inhibitor,

$$v = k_3\,[\mathrm{ES}]$$

so that, substituting for [ES] from equation (3.3), we get

$$v = \frac{k_3 e}{1 + \left(\frac{k_2 + k_3}{k_1[\mathrm{S}]}\right)\left(1 + \frac{[\mathrm{I}]}{K_i}\right)}$$

$$= \frac{k_3 e[\mathrm{S}]}{[\mathrm{S}] + \frac{(k_2 + k_3)}{k_1}\left(1 + \frac{[\mathrm{I}]}{K_i}\right)} \tag{3.4}$$

$$\therefore v = \frac{V_{max}\,[\mathrm{S}]}{[\mathrm{S}] + K_m\left(1 + \frac{[\mathrm{I}]}{K_i}\right)} \tag{3.5}$$

Equation (3.4) should be compared with equation (2.8) on p. 21, and equation (3.5) with equation (2.5) on p. 17. If we follow through from the enzyme conservation equation to see where the separate parts of equations (3.4 and 3.5) come from, it is clear that [S] in the denominator again represents the fraction of the enzyme present as ES. As [S] is raised and ES begins to predominate over E and EI, the second term in the denominator becomes insignificant. Since only this second term contains, we are left with the unchanged maximum rate, $k_3 e$, as predicted above.

The right half of the denominator, however, is the K_m term, proportional to the contribution of all forms of the enzyme without substrate bound. This is no longer just $(k_2 + k_3)/k_1$ as in equation (2.8), but is now multiplied by $1 + [\mathrm{I}]/K_i$. The balance between the 1 and the $[\mathrm{I}]/K_i$ represents the balance between E and EI. The increase in apparent K_m agrees again with our qualitative prediction. If $[\mathrm{I}] = K_i$, then $1 + [\mathrm{I}]/K_i = 2$, so that the K_m is exactly doubled. Note that this does *not* mean that the rate is halved: that would be true only with very low [S], whereas, at the other extreme, high [S], such a concentration of inhibitor would have no effect.

The graphical consequences of competitive inhibition are shown in Fig. 3.3. Lineweaver–Burk plots for different inhibitor concentrations give a family of lines fanning out from the same ordinate intercept since V_{max} is unaltered. Since this is a plot of $1/v$, addition of inhibitor makes the line *steeper*.

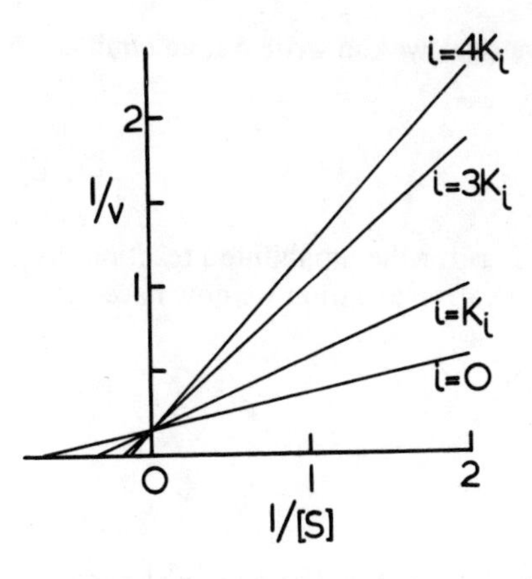

Fig. 3.3 Competitive inhibition as seen in the Lineweaver–Burk plot. Plots are shown for rates measured in the presence of a competitive inhibitor at several fixed concentrations, i, indicated in terms of the K_i against each line.

3.4 Uncompetitive inhibition

Consider next an inhibitor able to bind to ES but not to free enzyme, the precise opposite of our first case (Fig. 3.4). Even when S is saturating the enzyme is now still distributed between the product-forming complex ES and the inhibitor-containing complex EIS in a ratio that depends on [I] and the value of K_i. If, therefore, EIS is a 'dead-end' complex, unable to yield products directly, we may predict that V_{max} will be decreased by the factor $1 + [I]/K_i$. What about the K_m? K_m is essentially a measure of how hard you have to push with substrate to get enzyme from the substrate-free state to the substrate-bound state. Since I combines only with ES, its presence actually pulls enzyme over into the substrate-bound state. This means paradoxically that the inhibitor *lowers* the K_m! This does not however make it an activator. The rate approaches its ceiling faster than without I, but that ceiling is lower. There is no value of [S] for which the rate is higher in the presence of I than in its absence.

The case we have just considered is known as *uncompetitive* inhibition. Let us again immediately check the qualitative predictions by working

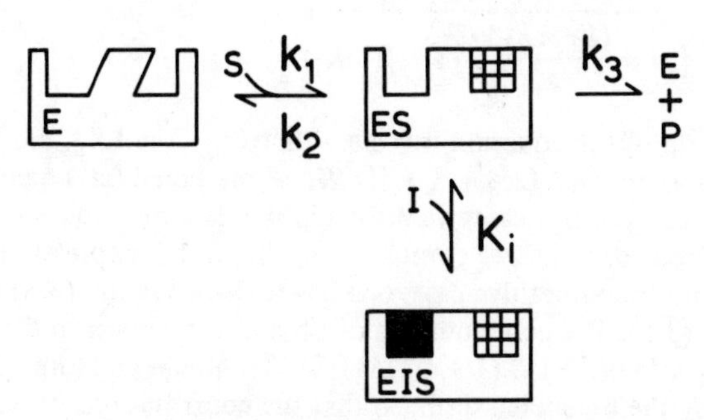

Fig. 3.4 Uncompetitive inhibition.

out the algebra. As in the competitive case, we can write out an equilibrium equation for [EIS] :–

$$[\mathrm{EIS}] = \frac{[\mathrm{ES}][\mathrm{I}]}{K_\mathrm{i}} \tag{3.6}$$

Once again the steady-state is defined, as for the uninhibited reaction, by equation (3.2). For the enzyme conservation equation we now have

$$e = [\mathrm{ES}] + [\mathrm{EIS}] + [\mathrm{E}]$$

Substituting from equation (3.6),

$$e = [\mathrm{ES}]\left(1 + \frac{[\mathrm{I}]}{K_\mathrm{i}}\right) + [\mathrm{E}]$$

Bringing in the steady-state equation (3.2) relating [E] and [ES] we obtain

$$e = [\mathrm{ES}]\left(1 + \frac{[\mathrm{I}]}{K_\mathrm{i}}\right) + [\mathrm{ES}]\frac{(k_2 + k_3)}{k_1[\mathrm{S}]}$$

$$= [\mathrm{ES}]\left[\left(1 + \frac{[\mathrm{I}]}{K_\mathrm{i}}\right) + \frac{(k_2 + k_3)}{k_1[\mathrm{S}]}\right] \tag{3.7}$$

As before, $v = k_3[\mathrm{ES}]$. Using equation (3.7), therefore,

$$v = \frac{k_3 e}{\left(1 + \frac{[\mathrm{I}]}{K_\mathrm{i}}\right) + \frac{(k_2 + k_3)}{k_1[\mathrm{S}]}} = \frac{k_3 e[\mathrm{S}]}{\left(1 + \frac{[\mathrm{I}]}{K_\mathrm{i}}\right)[\mathrm{S}] + \frac{(k_2 + k_3)}{k_1}}. \tag{3.8}$$

This is not yet in the familiar form of the Michaelis–Menten equation, however. If we want the second half of the denominator to give the apparent K_m in the presence of inhibitor, the first half must be [S]. We therefore divide through top and bottom by $1 + [\mathrm{I}]/K_\mathrm{i}$ to obtain equation (3.9):

$$v = \frac{k_3 e[\mathrm{S}]/(1 + [\mathrm{I}]/K_\mathrm{i})}{[\mathrm{S}] + \frac{(k_2 + k_3)}{k_1}\Big/(1 + [\mathrm{I}]/K_\mathrm{i})}. \tag{3.9}$$

In this form of the equation it is clear that V_{max} and K_m are both reduced by the same factor, $1 + [\mathrm{I}]/K_\mathrm{i}$, as predicted (cf. equation (2.8)).

In the competitive case, we dissected the denominator, identifying the contributions of the three complexes to the overall expression. To do this for the uncompetitive case, one has to use equation (3.8) rather than equation (3.9). The contributions of ES and EIS appear in the denominator of equation (3.8) as $(1 + [\mathrm{I}]/K_\mathrm{i})[\mathrm{S}]$. To obtain equation (3.9), we divided by the bracketed term, so that the contributions of both these complexes appear together in the denominator simply as [S].

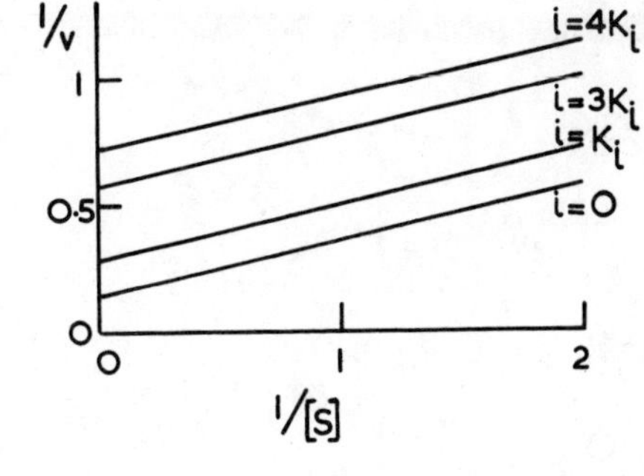

Fig. 3.5 Uncompetitive inhibition as seen in the Lineweaver–Burk plot.

Uncompetitive inhibition is easily recognized from Lineweaver–Burk plots. Since the ordinate intercept gives $1/V_{max}$ and the abscissa intercept gives $-1/K_m$, an uncompetitive inhibitor increases both intercepts by the same factor, giving a family of parallel lines (Fig. 3.5) for different values of [I].

3.5 Non-competitive inhibition

The final pattern we shall consider is that of *simple non-competitive* inhibition. (There is no logical basis for the distinction between 'un' and 'non'; one just has to remember it.) A non-competitive inhibitor can bind to both E and ES to form EI and EIS (Fig. 3.6). It is assumed that EIS cannot break down to give product. The further categorization of an inhibitor as *simple* non-competitive indicates that the K_i's for formation of the two inhibitor-containing complexes are equal i.e. that substrate does not affect the ease with which the enzyme binds the inhibitor. If [S] is saturating, E and EI drop out of the picture leaving a situation similar to the uncompetitive case considered above. Thus V_{max} must be decreased by the factor $1 + [I]/K_i$. In the uncompetitive case I also lowered the K_m by 'pulling' on ES. Here, however, I pulls equally on E and ES and therefore does not affect the distribution between them. The K_m should accordingly be unaltered by the inhibitor. In the Lineweaver–

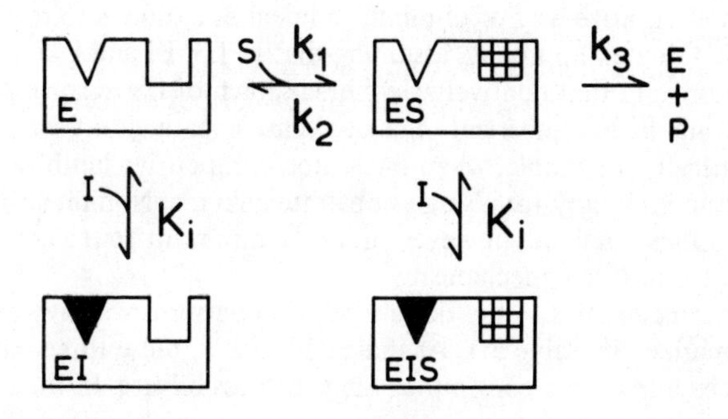

Fig. 3.6 Non-competitive inhibition.

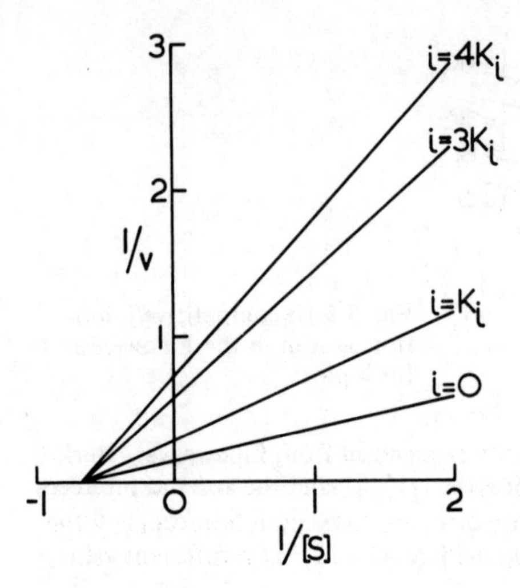

Fig. 3.7 Non-competitive inhibition as seen in the Lineweaver–Burk plot.

Burk plot (Fig. 3.7) different concentrations of a simple non-competitive inhibitor lead to a family of lines fanning out from a common point of intersection ($-1/K_m$) on the *abscissa.* It should be a simple exercise to derive the corresponding equation (3.10).

$$v = \frac{k_3 e[\mathrm{S}]/(1 + [\mathrm{I}]/K_i)}{[\mathrm{S}] + \left(\dfrac{k_2 + k_3}{k_1}\right)}. \tag{3.10}$$

In Fig. 3.6 there is no reaction arrow directly linking EI and EIS. The explanation is one of mathematical expediency. The arbitrary assumption that direct interconversion of EI and EIS either does not occur or else is 'kinetically insignificant' (i.e. slow!) enables us to retain the Briggs–Haldane steady-state treatment without getting into deep water. The extra reaction would introduce a cyclic pathway, and steady-state analysis would then lead to an equation containing terms in $[\mathrm{S}]^2$ and $[\mathrm{I}]^2$. An alternative way of obtaining a linear equation is to revert to the Michaelis–Menten approach, assuming that E, ES, EI and EIS are all at equilibrium, and that relatively slow breakdown of ES to form product is rate-limiting. Either linearization is in a sense cheating, but such devices are empirically justifiable, when linear non-competitive inhibition is found experimentally for a single-substrate enzyme. Non-linear patterns are sometimes obtained, however, and it is important to realize that they can arise from simple mechanisms.

The characteristics of the three inhibition patterns we have examined are summarized in Table 3.1. As indicated earlier, these three patterns are extreme examples in a continuous range of possibilities. In the non-competitive case, for instance, the two K_i values might differ. If so, the Lineweaver–Burk plots would intersect either above or below the

Table 3.1 Characteristics of different types of reversible enzyme inhibition

Type of inhibition	*Inhibitor combines with:–*	*Effect on* V_{max}	*Effect on* K_m	*Effect on* $1/v$ *against* $1/[S]$ *plot*
Competitive	E	Unchanged	Increased	Convergence on ordinate axis
Uncompetitive	ES	Decreased	Decreased	Parallel lines
Simple non-competitive	E and ES	Decreased	Unchanged	Convergence on abscissa axis.

abscissa. This pattern is often met, and is called 'mixed' inhibition by some authors and 'non-competitive' (without the 'simple') by others. In view of the wide range of possibilities, the most important thing is to have a qualitative grasp of *how* different inhibition patterns arise, and the confidence to work out new patterns from first principles. It is less important to memorize formulae for special cases.

3.6 Determination of K_i's

K_i values provide a useful yardstick of the relative potency of different inhibitors. The simplest way to determine a K_i would be to measure initial rates for a series of substrate concentrations with and without inhibitor added at a single fixed concentration. As we have seen above, in the Lineweaver–Burk plot the slope, ordinate intercept, or both may be altered by the factor $1 + [I]/K_i$. It should therefore be a simple matter to obtain K_i by substituting the known value of [I]. This method, based on one inhibitor concentration only, is sound but not very accurate in practice, and would not detect *non-linear* inhibition. It is better to use several inhibitor concentrations. K_i can then be obtained graphically rather than by calculation. For instance, for competitive inhibition,

$$\text{SLOPE} = (\text{UNINHIBITED SLOPE})(1 + [I]/K_i).$$

A plot of SLOPE against [I] should therefore give a straight line with a negative abscissa intercept equal to K_i (Fig. 3.8).

Dixon [32] proposed an alternative procedure: instead of plotting $1/v$ against $1/[S]$ and then replotting slopes or intercepts against [I], Dixon plotted $1/v$ directly against [I]. A replot, this time against $1/[S]$, would again yield K_i, but the merit of the Dixon plot is that K_i can in fact be estimated *without* replotting. Consider the competitive case again. Writing equation (3.5) in the reciprocal form:

$$1/v = 1/V_{max} + (1 + [I]/K_i)K_m/V_{max}[S] \tag{3.11}$$

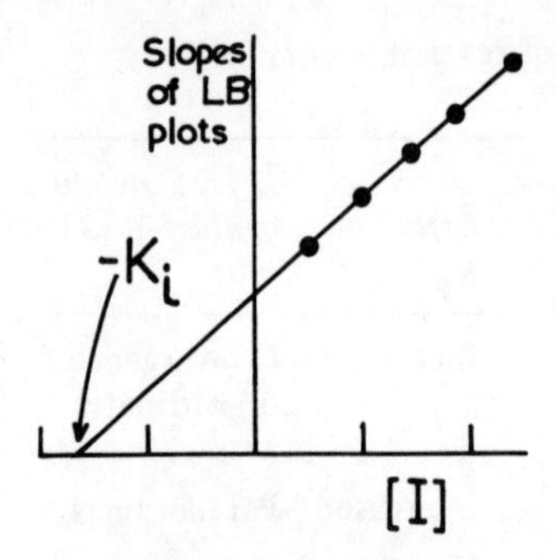

Fig. 3.8 Determination of the K_i of a competitive inhibitor from a secondary replot of the slopes of Lineweaver–Burk plots for various inhibitor concentrations.

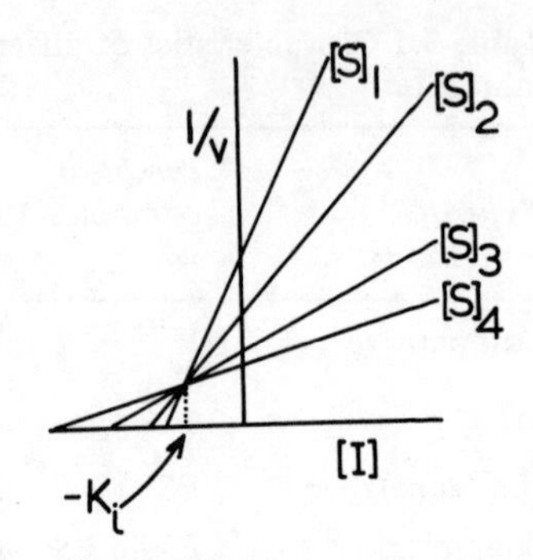

Fig. 3.9 The Dixon plot for determination of the K_i for a competitive inhibitor.

A plot of $1/v$ against [I] for a fixed value of [S] gives a straight line. If this is repeated for different values of [S], the lines all intersect at the same point in the fourth quadrant (Fig. 3.9). At the intersection for two substrate concentrations S_1 and S_2, we know from equation (3.11) that

$$1/V_{max} + (1 + [I]/K_i)K_m/V_{max}\,S_1$$
$$= 1/V_{max} + (1 + [I]/K_i)K_m/V_{max}S_2.$$

Most of this cancels, leaving

$$\frac{1}{S_1}\left(1 + \frac{[I]}{K_i}\right) = \frac{1}{S_2}\left(1 + \frac{[I]}{K_i}\right)$$

This equation can only be true either if $S_1 = S_2$, which we know is not the case, or if $1 + [I]/K_i = 0$, leading to $[I] = -K_i$. Thus, whatever the values of [S], the lines will intersect at a point where $[I] = -K_i$, allowing the value of K_i to be read straight off the Dixon plot.

In the case of simple non-competitive inhibition this procedure will also work, and the intersection is on the abscissa. For *un*competitive inhibition, however, the Dixon plot, like the Lineweaver–Burk plot, gives parallel lines, as you may easily verify. In this case, therefore, one has to fall back on a re-plot of the ordinate intercepts, and the Dixon plot offers no advantage.

4 pH effects

4.1 Introduction

Enzymes are complex structures, and, accordingly the effects of pH variation on enzyme-catalysed reactions are often also complex. Nevertheless, under favourable circumstances, pH studies can yield much useful information. Several types of pH effect may be distinguished, as considered below.

4.2 Position of equilibrium

Many reactions involve the net uptake or release of protons. Clearly for such reactions an alteration in the buffered pH shifts the position of equilibrium. For example, reaction (8), catalysed by glutamate dehydrogenase, has an equilibrium constant very unfavourable for glutamate oxidation at pH 7 with reactant concentrations in a reasonable range.

$$\text{L-glutamate} + NAD^+ + H_2O \rightleftharpoons \text{2-oxoglutarate} + NADH + NH_3 + H^+ \qquad (8)$$

Buffering at a higher pH effectively traps the proton produced in the reaction. At pH 9, the H^+ concentration is 100 times lower than at pH 7, and glutamate oxidation can therefore proceed further. This, however, is opposite to the pH-dependence of the *rate*: the initial rate of oxidation of glutamate at pH 9 is lower than at pH 7. The reaction may not be able to get very far at pH 7, but it gets there quickly!

Many reactions may be driven to virtual completion in either direction by adjusting the pH, e.g. reaction (4), p. 22.

4.3 Instability

Proteins retain their native conformation within a limited pH range beyond which gross alterations occur. Such 'denaturation' is reversible in theory, and sometimes in practice, but reversal is usually slow by comparison with the rate of the enzyme-catalysed reaction. For our present purposes denaturation should be regarded as a depletion of enzyme concentration distinct from the reversible effects considered under Section 4.5 below. The pH stability varies greatly from one enzyme to another.

4.4 Substrate ionizations

Some substrates are uncharged (e.g. glucose, ethanol), but the majority are capable of acquiring a charge within the normal pH range. Carboxyl, amino and phosphate groups, for example, are all charged at pH 7. The question then arises as to which of the ionic forms is the true substrate. If only one form of the substrate is active, the pH dependence of the rate should reflect the titration of the substrate, unless it is saturating. Substrate pK's may of course be measured separately without reference to the enzyme-catalysed reaction. For some enzymes a range of substrates may be used; this may offer the opportunity to explore the importance of the state of ionization of an individual group.

4.5 Ionization of 'essential' residues

Within the structure of an enzyme molecule, some residues play a more crucial role than others. They may be important in maintaining the active conformation of the enzyme. Equally, they may control access to, or form part of, the active site. Specific chemical modification is often used to identify such residues, but, since many of them are ionizable, pH studies can also yield useful clues. (It is also possible to combine these two approaches; the pH dependence of the alkylation of pancreatic ribonuclease [33, 34] by iodoacetate suggested two essential histidine residues, a result in agreement with indications from the pH dependence of of catalytic activity [35].)

The basis of such studies is that, even when instability, substrate ionizations etc. are accounted for, the rate of an enzyme-catalysed reaction is usually pH-dependent. The shape of the pH profile varies: there is often a broad plateau, with a fall-off apparent sometimes on one side only (Fig. 4.1a) and sometimes on both (Fig. 4.1b). In the plateau region the enzyme is all in its active form. If, however, the pK's of the essential groups are closer than in Fig. 4.1b, a bell-shaped profile may result (Fig. 4.1c). (The pK of a group is minus the logarithm of its ionization constant; a group with an ionization constant of 10^{-9} M has a pK of 9 and is 50% ionized at pH 9.) This means that, with increasing pH, before one group has been fully de-protonated to give its active form, another group required in the protonated form has started to lose its proton.

Now, even for a one-substrate reaction, the initial rate is a function of V_{max}, K_m and [S]. Since K_m and V_{max} may show dissimilar dependences on pH, the pH dependence of the rate for sub-saturating [S] is likely to be complex. If pH studies are to shed any light on the nature of essential ionizable residues, the separate effects on K_m and V_{max} must be determined. Shortcuts are risky: one cannot assume, for example, that, because 1 mM substrate is saturating at pH 6, it will still be saturating at pH 8. Proper kinetic analysis at each pH is essential.

In considering the analysis of pH curves, we shall concentrate on the variation of V_{max} with pH. (A basically similar, though more complex, treatment can be applied to the variation of K_m.) The analysis [36] rests on the simplifying assumption that each apparent pK discernible in the profile represents the ionization of a single group in the enzyme [37]. In fact there are bound to be other groups ionizing over the same pH range

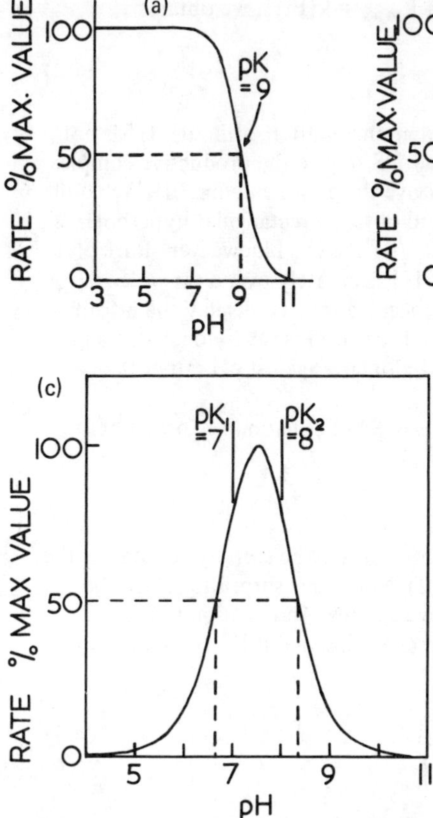

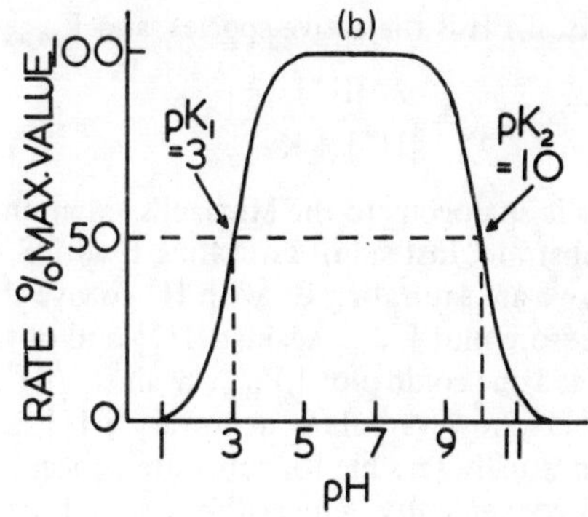

Fig. 4.1 pH dependence of the rates of enzyme-catalysed reactions. The curve in (a) reflects the ionization of a single essential residue required in the protonated state and possessing a pK of 9. In (b) two ionisations are apparent, one group with a pK of 3 being required in the deprotonated state and another with a pK of 10 being required in the protonated state. In (c) there are again two pK's but this time much closer, 7 and 8.

as the postulated essential group. One has to assume that their effect on the rate is negligible.

If the ionization constant for the residue of interest is K, we have:

$$\frac{[E^-][H^+]}{[EH]} = K \tag{4.1}$$

where E^- and EH are heterogeneous with respect to the state of ionization at other positions. As usual, we also have an enzyme conservation equation:

$$e = [EH] + [E^-]$$

Substituting from equation (4.1), therefore,

$$e = [EH] + \frac{[EH]K}{[H^+]} = [EH]\left(1 + \frac{K}{[H^+]}\right)$$

$$\therefore [EH] = \frac{e}{(1 + K/[H^+])} = \frac{e[H^+]}{[H^+] + K}.$$

Thus, if EH is the active species, and $V_{max} = k[EH]$, we obtain:

$$V_{max} = \frac{ke[H^+]}{[H^+] + K}. \qquad (4.2)$$

This is analogous to the Michaelis–Menten equation, although H^+ is not a substrate. Instead of saturating E with S to give the productive complex ES, we are saturating E^- with H^+ to give the active enzyme EH. We could therefore plot V_{max} against $[H^+]$ and obtain a rectangular hyperbola. Indeed one could plot $1/V_{max}$ against $1/[H^+]$ as in a Lineweaver–Burk plot. Buffers, however, allow us to vary $[H^+]$ accurately over a far wider range than usually feasible for substrate concentration. To display the information conveniently, a logarithmic scale is used. This is why one normally encounters, as in Fig. 4.1a, a sigmoid plot of v against pH rather than a hyperbolic plot against $[H^+]$.

If E^- is the active enzyme form, instead of equation (4.2) we obtain:

$$V_{max} = \frac{keK}{[H^+] + K}.$$

This is not in a Michaelis–Menten form, as may be seen by comparing the numerator with that in equation (4.2). Nor is this surprising, since the enzyme *releases* H^+ to form the active species. An equation in the Michaelis–Menten form can, however, be obtained if H^+ is replaced by $10^{-14}\,M^2/[OH^-]$, giving:

$$V_{max} = \frac{ke[OH^-]}{[OH^-] + \left(\frac{10^{-14}M^2}{K}\right)}. \qquad (4.3)$$

The corresponding pH profile is another sigmoid, but showing this time an increase in rate with increasing pH.

Regardless of whether the protonated or the unprotonated species is the active one, the pK for the ionization is the pH at the point of inflexion of the sigmoid curve (Figs. 4.1a,b). It may also be obtained from a plot of log V_{max} against pH as follows. Taking the case in which EH is active, equation (4.2) may be rewritten:

$$V_{max} = \frac{V_{max(true)}}{1 + \frac{K}{[H^+]}}$$

$$\therefore \log V_{max} = \log V_{max(true)} - \log\left(1 + \frac{K}{[H^+]}\right). \qquad (4.4)$$

In this equation V_{max} is the experimental value measured at any given pH, whereas $V_{max(true)}$ is the value that is or would be attained with all the enzyme present in the active form.

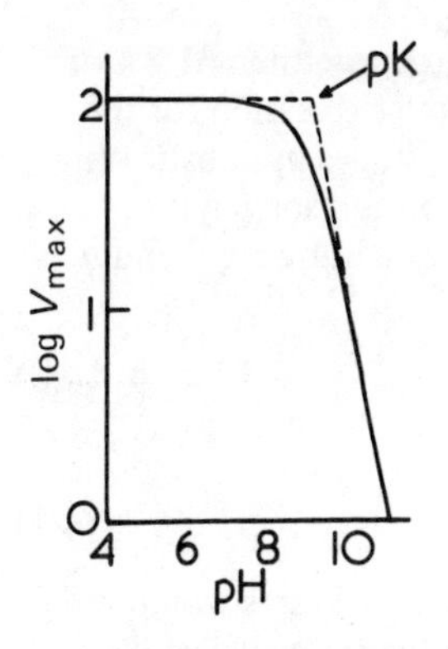

Fig. 4.2 Determination of pK (in this case 9) from a plot of log V_{max} against pH.

Now, when $[H^+] \gg K$, equation (4.4) reduces to:

$$\log V_{max} = \log V_{max(true)} - \log 1 = \log V_{max(true)}.$$

Conversely, when $[H^+] \ll K$ (i.e. $pH \gg pK$),

$$\log V_{max} = \log V_{max(true)} - \log K + \log[H^+].$$

A plot of log V_{max} against pH therefore gives a curve (Fig. 4.2) that is asymptotic to two straight lines, one horizontal at an ordinate of log $V_{max(true)}$, and one with a slope of –1. That the lines intersect at pH = pK can be shown by solving the two equations. When the activity drops off at acid pH, the mirror image plot is obtained, with a slope of +1 at acid pH, a horizontal line above the pK, and again intersection at the pK.

This procedure can be extended to deal with two ionizations. If one group has to be protonated and one not, the log plot gives lines of slope +1 and –1 separated by a horizontal section, and both pK's can be obtained as before. If, however, the pK's are less than two pH units apart, the two protonation steps overlap so that the protein is never completely in its active form. The plateau disappears, and the midpoints of the ascending and descending slopes of the resulting bell-shaped curve are displaced from the true pK values (Fig. 4.1c). It is nevertheless possible to calculate the pK's from such a curve, as Alberty and Massey showed [38]. If EH, the active species, is converted with decreasing pH to EH_2^+ with an ionization constant K_1, and with increasing pH to E^- with an ionization constant K_2, then the fraction, F, of enzyme in the active form is given by equation (4.5):

$$F = \frac{[EH]}{e} = \frac{[H^+]}{K_2 + [H^+] + \dfrac{[H^+]^2}{K_1}}. \tag{4.5}$$

Differentiation with respect to $[H^+]$ shows that F reaches a maximum of $K_1/[K_1 + 2\sqrt{(K_1K_2)}]$ when $[H^+] = \sqrt{(K_1K_2)}$. Thus, for example, if the two ionisable groups have pK's of 7 and 8 ($K_1 = 10^{-7}$ M, $K_2 = 10^{-8}$ M), F

attains a maximal value of 0.613 at pH 7.5, i.e. no more than 61.3% of the enzyme is ever in the active form. Knowing the expression for the maximum, we can now find the midpoints of the slopes by substituting $F = 0.5\, K_1/(K_1 + 2\sqrt{(K_1 K_2)})$ into equation (4.5) and solving the resulting quadratic for $[H^+]$. If the roots of this quadratic are p and q, then

$$pq = K_2 K_1 \tag{4.6}$$

and

$$p + q = K_1 + 4\sqrt{(K_1 K_2)} \tag{4.7}$$

With true pK's of 7 and 8 (Fig. 4.1c), the half-maximal rates would be found experimentally at pH 6.65 and 8.35 (note the symmetrical displacement from the true pK's), so that $p = 2.22 \times 10^{-7}$ M (antilog –6.65) and $q = 4.5 \times 10^{-9}$ M (antilog –8.35). Hence pq = 10^{-15} M^2 and $p + q = 2.265 \times 10^{-7}$ M. Using equation (4.6) and substituting in equation (4.7), one would obtain

$$K_1 = 2.265 \times 10^{-7}\ \text{M} - 4 \times 10^{-7.5}\ \text{M}$$
$$= 2.265 \times 10^{-7}\ \text{M} - 1.265 \times 10^{-7}\ \text{M} = 10^{-7}\ \text{M}$$

Hence $K_2 = 10^{-8}$ M.

Dixon and Webb [39] emphasize the need to use K's related to the ionization of particular groups rather than to the overall charge state of the protein. As they point out, if the relevant pK's are similar, there are bound to be two forms of EH, the active form and another with the proton on the 'wrong' group. This does not affect the procedure for determining K_1 and K_2 from the midpoints of the slopes, but it does mean, for instance, that, if $K_1 = K_2$ and if the two ionizations do not affect one another, the maximum possible fraction of the enzyme in the active form is 1/4, not 1/3 as predicted by equation (4.5). The appropriately amended form of equation (4.5) would include a term in the denominator representing the contribution of the inactive form of EH.

Sometimes two groups of the same charge may be required. In this case, if the pK's are sufficiently separated, one obtains a horizontal line and further linear sections with slopes of 1 and 2, or –1 and –2, in the log plot.

Various limitations in the analytical approach have been indicated above. There are two other potential complications:–

(i) if the ionization does not result in complete loss of activity, the plot of log V_{max} against pH will not be linear;

(ii) a kinetic constant such as V_{max} may be a composite function of several rate constants.

Unless all these drop to zero upon protonation or de-protonation of the the relevant residue, the measured pK may be displaced from the true value. For example, on p. 62 we have a 2-substrate rate equation in which the maximum rate is a function of two rate constants. Let us suppose that one of these (k_{ind}) is pH independent over the range investigated, but that the other (k_{dep}) is governed by the state of ionization of an essential residue which has to be protonated for catalysis to occur, and has a pK of 7. At pH 5 this group will be almost entirely protonated, so that k_{dep} will have its maximal value. If, even at this pH, k_{dep} is largely rate-limiting (e.g. $k_{dep} = 0.01 \times k_{ind}$), then, as the pH is raised, V_{max} will drop in parallel with the value of k_{dep}. Let us, however, consider the opposite situation, in which, at pH 5, $k_{dep} = 100 \times k_{ind}$. If the pH is raised past the pK so that k_{dep} is decreased by a factor of 10, V_{max} is very little affected, because k_{ind}, still ten times smaller than k_{dep}, remains the major factor determining the rate. Even if k_{dep} is decreased 100-fold, V_{max} only decreases by a factor of two. Only when k_{dep} is decreased 1000-fold or more does it take over as the rate-limiting constant, so that V_{max} decreases 10-fold for each unit rise in pH. The apparent pK from a plot of log V_{max} against pH would thus be a full two units higher than the true value.

These cautionary comments apply, if anything, with even greater force to studies of the pH dependence of K_m. V_{max} sometimes reflects a single rate constant, but K_m is always a function of at least two rate constants.

Despite these reservations, there are sometimes good reasons for believing that a measured kinetic parameter reflects the value of a single rate constant (see also Chapter 5), and that the apparent pK is a true one. Once this is established, the next step is to consider the likely chemical nature of the ionizable group. This may lead to chemical modification studies designed to confirm the identification. It may be possible to isolate a modified peptide, and place the essential residue within the overall amino acid sequence, if this is known. One may then tackle the next problem, which is to establish *why* the group is essential. pH studies thus form only a small part of what may be a lengthy programme.

The provisional identification of an essential residue from its pK is based on Table 4.1 which lists the normal range of pK's for ionizable

Table 4.1 The normal pK range of ionizable groups in proteins

Ionizable group	*pK range*
α-COOH	3.0–3.2
aspartyl –COOH	3.0–4.7
glutamyl –COOH	4.4 (approx.)
histidine imidazolium	5.6–7.0
α–NH_2	7.6–8.4
–SH	8.3–8.6
lysyl ϵ–NH_2	9.4–10.6
tyrosyl –OH	9.8–10.4
arginine guanidinium	11.6–12.6

groups in proteins. Even here a potential pitfall exists. An essential residue may owe its functional importance to the very fact that its ionization constant is atypical! Thus, although the normal pK range for lysine $-NH_2$ groups is 9.4–10.6, glutamate dehydrogenase has an essential lysine with a pK of 8. In summary, therefore, pH studies, while often very useful, can easily lead to erroneous conclusions. They should be interpreted with caution.

5 More than one substrate

5.1 Multi-substrate enzymes

Up to now we have considered one-substrate enzymes only. In fact, the only true one-substrate enzymes are the isomerases, which catalyse reactions of the type A $\rightleftharpoons$ B, and the lyases, which catalyse reactions of the type A $\longrightarrow$ B + C, and are therefore one-substrate enzymes in one direction only. Hydrolases, catalysing reactions of the general type

$$A\text{–}B + H_2O \rightleftharpoons A\text{–}OH + BH$$

may be regarded as honorary one-substrate enzymes, since one substrate, water, is normally present at constant concentration.

The remaining groups are:

(i) oxidoreductases:

$$\text{oxidant} + \text{reductant} \rightleftharpoons \text{reduced product} + \text{oxidized product}$$

(ii) transferases:

$$A + BX \rightleftharpoons AX + B$$

This large group includes methyl transferases, transaldolase, transketolase, acyl transferases, glycosyl transferases, transaminases, kinases etc.

(iii) ligases:

$$X + Y + ATP \rightleftharpoons XY + ADP + P_i \text{ (or } AMP + PP_i)$$

These three groups of enzymes by definition catalyse reactions involving more than one substrate.

5.2 Cofactors

Before proceeding with the kinetics of multi-substrate enzymes, we should deal with a potential source of semantic confusion, namely the distinction between substrates, coenzymes and prosthetic groups. Many enzymes employ either metal ion cofactors. (Mg^{++}, Fe^{++}, Cu^{++}, Mo^{+++}, Zn^{++}) or organic cofactors. Thus many oxidoreductases utilize haem, nicotinamide or flavin; the transferases use folate, coenzyme A, pyridoxal phosphate, thiamine and adenosine phosphates; some of the ligases use biotin etc. Among these cofactors, some may be regarded as integral parts of their enzymes; they are known as *prosthetic* groups. Others are substrates in that they combine with the enzyme and leave it again in the course of a single catalytic cycle; these are called *coenzymes*. This distinction is illustrated by the roles of NAD^+ and FAD in the oxidation of an amino acid ($R \cdot CHNH_2 \cdot COOH$) to the corresponding oxoacid ($R \cdot CO \cdot COOH$) and NH_3. L-amino acid oxidase is a flavoprotein; reducing equivalents from the amino acid are received by the tightly bound FAD prosthetic group, which then remains reduced until an oxidising substrate, in this case O_2, appears. The FAD is then reoxidised, and remains bound, as before, while the H_2O_2 departs. Consider, by contrast, the amino acid oxidation (reaction 8, p. 37) catalysed by glutamate dehydrogenase. This enzyme is not a flavoprotein, and cannot catalyse oxidation of the glutamate until a second, oxidizing substrate arrives. That second substrate is the coenzyme, NAD^+. Having received the reducing equivalents, rather than waiting on the enzyme surface, like FAD, for another oxidant, the coenzyme leaves to be oxidized elsewhere by any one of a number of other enzyme systems. With the flavoprotein there are two redox reactions, and the cofactor acts in an intermediary capacity, passing on the reducing equivalents from an external donor to an external acceptor. In the glutamate dehydrogenase reaction the cofactor *is* the external acceptor, and only one redox reaction is involved. In effect, enzymes with prosthetic groups may be thought of as two enzymes rolled into one, with the added sophistication that the 'middle' substrate never comes off.

Haem, biotin and pyridoxal phosphate, like flavin, usually function as prosthetic groups, whereas tetrahydrofolate, Coenzyme A and adenosine phosphates, like $NAD(P)^+$, act as coenzymes. Most metal ion cofactors serve as prosthetic groups. Mg^{++} is an exception in that, with many kinases, the metal ion does come on and off during the catalytic cycle.

Although a coenzyme as defined above is clearly a substrate, the word 'substrate' is nevertheless frequently used loosely to distinguish the non-coenzyme substrate. Thus, in reaction 8 (p. 37), glutamate might be referred to as 'the substrate' and NAD^+ as 'the coenzyme'. This usage has no mechanistic basis and is a metabolic distinction. 'Substrates' in this sense are the raw material of metabolic pathways, whereas coenzymes are auxiliary agents linking those pathways.

5.3 Two-substrate reaction pathways

For a one-substrate reaction, there can be no argument about the order of addition of reactants. With more reactants choices arise. Consider the possibilities for a reaction $A + B \rightleftharpoons P + Q$:

(i) Reaction may not occur until both substrates are bound at the active site forming a *ternary complex*, EAB. This is a *sequential* mechanism. This still leaves open the question of the routes to and from EAB. If B cannot bind to free E, but only to the binary complex EA, and, symmetrically, Q can only bind to EP, then we have a *compulsory-order* mechanism:

$$E \rightleftharpoons EA \rightleftharpoons EAB \rightleftharpoons EPQ \rightleftharpoons EP \rightleftharpoons E.$$

One could equally well envisage EA and EQ, or EB and EP, or EB and EQ as the allowed pairs of binary complexes. These possibilities are completely equivalent as long as A, B, P and Q are only theoretical substrates of a theoretical enzyme, but they become distinct as soon as the symbols are assigned to the reactants of a real enzyme.

A third possibility is that the enzyme may follow *either* pathway to the central complex, the ratio of fluxes through the two paths depending on the actual substrate concentrations. This is called a *random-order* mechanism:

$$\begin{array}{ccccccc} & \mathrm{EA} & & & & \mathrm{EP} & \\ \mathrm{E} & & \mathrm{EAB} & \rightleftharpoons & \mathrm{EPQ} & & \mathrm{E} \\ & \mathrm{EB} & & & & \mathrm{EQ} & \end{array}$$

(ii) It may not be necessary for both substrates to be present simultaneously on the enzyme surface. Indeed it may not even be possible, because in some cases the two substrates share the same site e.g. transaminases (reaction 2, p. 11). If the substrates do not meet at the active site, clearly one substrate must leave something on the enzyme to be passed on to the other substrate. Thus, if we replace the substrate symbols A, B, P and Q by AX, B, A and BX, we may represent the mechanism as shown below:

$$\begin{array}{ccccccccc} \mathrm{E} & \longrightarrow & \mathrm{EAX} & \longrightarrow & \mathrm{EX} & \longrightarrow & \mathrm{EBX} & \longrightarrow & \mathrm{E} \\ & \mathrm{AX} & & \mathrm{A} & & \mathrm{B} & & \mathrm{BX} & \end{array}$$

(X may denote reducing equivalents, a phosphate group, an amino group, an acyl group etc.) For obvious reasons this is known as an *enzyme substitution* mechanism. The enzyme exists in two 'free' (i.e. devoid of A or B) forms, E and the chemically substituted form EX, which is often sufficiently stable to be isolated. Such a mechanism may also be referred to as a *double displacement* mechanism, by analogy with ionic reactions, or as a *ping-pong* mechanism. Enzymes with prosthetic groups often follow mechanisms of this type.

5.4 Objectives

In analysing the order of substrate addition some kineticists aim only to provide an operational description of the enzyme as it functions within limited ranges of substrate concentration. Others seek to go further and define the capabilities of their enzymes in more absolute terms. The author once heard a kineticist, in defending a compulsory-order mechanism for his enzyme, refer to 'an insignificant leakage via alternative pathways at very low coenzyme concentrations'. The 'low' concentrations were in fact physiological, but, even if they had not been, the very existence of alternative pathways tells us something important about the active centre of the enzyme, namely that it is able to bind each substrate in the absence of the other.

In the case of a *true* compulsory-order mechanism, one has to consider why the free enzyme is able to bind only one of its substrates.

Lactate dehydrogenase, for instance, which catalyses reaction 9, follows a compulsory-order pathway with the coenzyme binding first in both directions of reaction [40].

$$CH_3 \cdot CHOH \cdot COOH + NAD^+ \rightleftharpoons CH_3 \cdot CO \cdot COOH + NADH + H^+ \quad \text{... (9)}$$

Is the coenzyme itself an indispensable part of the binding site for lactate/pyruvate? This seems hardly likely; NAD^+ serves as coenzyme for dozens of dehydrogenases which must differ by virtue of their specificity for the second substrate. The alternative is that binding of the coenzyme provokes a conformational change in the enzyme, making a latent substrate site fully functional. This prediction from the kinetics is convincingly borne out in the case of lactate dehydrogenase by X-ray crystallographic findings [41].

5.5 Two-substrate initial-rate equations

The kinetic analysis of an enzyme mechanism usually begins with an attempt to disentangle the possible reaction pathways by studying the dependence of the overall reaction rate on the substrate concentrations.

Once the stoicheiometry of reaction is established, the procedure, in a nutshell, is as follows:

(i) list all reasonable candidate mechanisms;

(ii) write down the corresponding initial-rate equations and note any distinctive features;

(iii) make a series of rate measurements with substrate concentrations systematically varied over a wide range;

(iv) analyse the results and compare the experimentally derived rate equation with those listed in (ii); draw conclusions;

(v) carry out any necessary confirmatory experiments.

This, then provides the context for the ensuing consideration of the derivation of initial-rate equations.

No matter how many substrates are involved, the basic method for deriving the rate equation is precisely that used earlier for the one-substrate case. The rate is determined by the distribution of the enzyme among its various complexes. One therefore uses the steady-state or equilibrium expressions relating the concentrations of these complexes in conjunction with the enzyme conservation equation. Having said this, we may either derive a general equation and simplify at the end by setting appropriate product concentrations to zero to give the initial-rate equation, *or* we may simplify at the outset, bypassing the general equation. The latter procedure gives simpler algebra and is adopted below.

5.6 Compulsory-order ternary-complex mechanism

If we make the initial-rate assumption, i.e. that rates are measured before products build up sufficiently to start reversing steps in the mechanism, the reaction scheme given on p. 46 may be simplified as follows:

$$\mathrm{E} \underset{k_2}{\overset{k_1}{\rightleftharpoons}} \mathrm{EA} \underset{k_4}{\overset{k_3}{\rightleftharpoons}} \mathrm{EAB} \underset{k_6}{\overset{k_5}{\rightleftharpoons}} \mathrm{EPQ} \xrightarrow{k_7} \mathrm{EP} \xrightarrow{k_9} \mathrm{E}.$$

By setting [P] and [Q] equal to zero we have eliminated the reverse steps from E to EP and from EP to EPQ. Since there are five enzyme-containing species, E, EA, EAB, EPQ and EP, we need four steady-state equations to solve for the distribution of the enzyme among these species. The concentrations of all the complexes are worked out below in terms of [EP]. This is convenient, since the rate is simply k_9 [EP], but other starting points could equally well be chosen.

(1) *Steady state for EP*:

$$[\mathrm{EP}]\, k_9 = [\mathrm{EPQ}]\, k_7$$

$$\therefore \quad [\mathrm{EPQ}] = [\mathrm{EP}]\, k_9/k_7. \qquad (5.1)$$

(2) *Steady state for EPQ*: $[\mathrm{EPQ}]\ (k_6 + k_7) = [\mathrm{EAB}]\ k_5$.

$\therefore$ Rearranging and substituting from equation (5.1),

$$[\mathrm{EAB}] = [\mathrm{EPQ}]\ \frac{(k_6 + k_7)}{k_5} = [\mathrm{EP}]\ \frac{k_9(k_6 + k_7)}{k_5 k_7}\ . \qquad (5.2)$$

(3) *Steady state for EAB*: $[\mathrm{EAB}]\ (k_4 + k_5) = [\mathrm{EPQ}]\ k_6 + [\mathrm{EA}]\ k_3[\mathrm{B}]$

$\therefore$ $[\mathrm{EA}]\ k_3\ [\mathrm{B}] = [\mathrm{EAB}]\ (k_4 + k_5) - [\mathrm{EPQ}]\ k_6$

$\therefore$ Using the expression for [EAB] in terms of [EPQ] (equation 5.2),

$$[\mathrm{EA}]\ k_3\ [\mathrm{B}] = [\mathrm{EPQ}]\ \left[\frac{(k_6 + k_7)(k_4 + k_5)}{k_5} - k_6\right]$$

$$= \frac{[\mathrm{EPQ}]}{k_5}\ (k_4k_6 + k_4k_7 + k_5k_7)$$

$$= \frac{[\mathrm{EP}]\, k_9}{k_5\, k_7}\ (k_4k_6 + k_4k_7 + k_5k_7)$$

$$\therefore \qquad [\mathrm{EA}] = \frac{[\mathrm{EP}]\ k_9\ (k_4k_6 + k_4k_7 + k_5k_7)}{k_3k_5k_7[\mathrm{B}]} \qquad (5.3)$$

The expression for [E] in terms of [EP] may now be obtained from either the steady-state equation for [EA] or that for [E]. The latter route is slightly simpler:

(4) *Steady state for E*: $[\mathrm{E}]\ k_1\ [\mathrm{A}] = [\mathrm{EP}]\ k_9 + [\mathrm{EA}]\, k_2$

$\therefore$ Using equation (5.3) for [EA] in terms of [EP],

$$[\mathrm{E}] = [\mathrm{EP}]\left[\frac{k_9}{k_1[\mathrm{A}]} + \frac{k_2k_9(k_4k_6 + k_4k_7 + k_5k_7)}{k_1k_3k_5k_7[\mathrm{A}]\,[\mathrm{B}]}\right]. \qquad (5.4)$$

We can now substitute in the enzyme conservation equation as follows:

(5) $e = [\mathrm{EP}] + [\mathrm{EPQ}] + [\mathrm{EAB}] + [\mathrm{EA}] + [\mathrm{E}]$

$$= [\mathrm{EP}]\left[1 + \frac{k_9}{k_7} + \frac{k_9(k_6 + k_7)}{k_5 k_7} + \frac{k_9(k_4 k_6 + k_4 k_7 + k_5 k_7)}{k_3 k_5 k_7 [\mathrm{B}]} + \frac{k_9}{k_1 [\mathrm{A}]} + \frac{k_2 k_9(k_4 k_6 + k_4 k_7 + k_5 k_7)}{k_1 k_3 k_5 k_7 [\mathrm{A}][\mathrm{B}]}\right]. \tag{5.5}$$

In the large bracket in equation (5.5), the first three terms, the contributions of EP, EPQ and EAB, are independent of substrate concentration. Then there is a term in $1/[\mathrm{B}]$, representing the contribution of EA. The last two terms, in $1/[\mathrm{A}]$ and $1/[\mathrm{A}][\mathrm{B}]$, are both contributed by the free enzyme, E.

Now $v = k_9[\mathrm{EP}]$. Dividing this equation into equation (5.5), we obtain the reciprocal form of the initial-rate equation, suitable for use in the Lineweaver-Burk plot:

$$\frac{e}{v} = \frac{1}{k_9} + \frac{1}{k_7} + \frac{(k_6 + k_7)}{k_5 k_7} + \frac{(k_4 k_6 + k_4 k_7 + k_5 k_7)}{k_3 k_5 k_7 [\mathrm{B}]} + \frac{1}{k_1 [\mathrm{A}]} + \frac{k_2(k_4 k_6 + k_4 k_7 + k_5 k_7)}{k_1 k_3 k_5 k_7 [\mathrm{A}][\mathrm{B}]}. \tag{5.6}$$

If we label the coefficients of the constant term and the terms in $1/[\mathrm{A}]$, $1/[\mathrm{B}]$ and $1/[\mathrm{A}][\mathrm{B}]$ as $\phi_0, \phi_\mathrm{A}, \phi_\mathrm{B}$ and ϕ_{AB} respectively, we may condense equation (5.6) as below:

$$\frac{e}{v} = \phi_0 + \frac{\phi_\mathrm{A}}{[\mathrm{A}]} + \frac{\phi_\mathrm{B}}{[\mathrm{B}]} + \frac{\phi_{\mathrm{AB}}}{[\mathrm{A}][\mathrm{B}]}. \tag{5.7}$$

Equation (5.7) emphasizes the algebraic form of the equation. The values of $\phi_0, \phi_\mathrm{A}, \phi_\mathrm{B}$ and ϕ_{AB} in terms of the rate constants in equation (5.6) are given in Table 5.1.

The equation reduces further at extremes of substrate concentration. Thus, if both [A] and [B] are very large, equation (5.7) reduces to:

$$\frac{e}{v} = \phi_0 \quad \therefore \quad \frac{V_{\max}}{e} = \frac{1}{\phi_0}. \tag{5.8}$$

If only B becomes saturating, then only the terms in $1/[\mathrm{B}]$ and $1/[\mathrm{A}][\mathrm{B}]$ drop out, leaving:–

$$\frac{e}{v} = \phi_0 + \frac{\phi_\mathrm{A}}{[\mathrm{A}]}$$

Hence, by rearrangement,

$$v = \frac{V_{\max}[\mathrm{A}]}{[\mathrm{A}] + \phi_\mathrm{A}/\phi_0}. \tag{5.9}$$

Table 5.1 Initial-rate kinetic parameters for the 2-substrate compulsory-order ternary-complex mechanism

The overall rate equation is equation (5.6). The ϕ parameters (Dalziel [42]) are the coefficients of the four terms in equation (5.7). The Alberty coefficients [43] given in the bottom half of the table are the coefficients of the equation in the form of equation (5.13).

$$\phi_0 = \frac{k_5k_7 + k_5k_9 + k_6k_9 + k_7k_9}{k_5k_7k_9} \qquad \phi_A = \frac{1}{k_1}$$

$$\phi_B = \frac{k_4k_6 + k_4k_7 + k_5k_7}{k_3k_5k_7} \qquad \phi_{AB} = \frac{k_2(k_4k_6 + k_4k_7 + k_5k_7)}{k_1k_3k_5k_7}$$

$$V_{max} = \frac{k_5k_7k_9e}{k_5k_7 + k_5k_9 + k_6k_9 + k_7k_9}$$

$$K_A = \frac{k_5k_7k_9}{k_1(k_5k_7 + k_5k_9 + k_6k_9 + k_7k_9)}$$

$$K_B = \frac{k_9(k_4k_6 + k_4k_7 + k_5k_7)}{k_3(k_5k_7 + k_5k_9 + k_6k_9 + k_7k_9)}$$

$$K_{AB} = \frac{k_2k_9(k_4k_6 + k_4k_7 + k_5k_7)}{k_1k_3(k_5k_7 + k_5k_9 + k_6k_9 + k_7k_9)}$$

This is the same in form as equation (2.5) (p. 17), the Michaelis–Menten equation for a one-substrate enzyme. Thus, if substrate B is saturating, the dependence of the rate on [A] follows the normal linear pattern, and, under these conditions, the K_m for A is ϕ_A/ϕ_0. Similarly, ϕ_B/ϕ_0 is the K_m for B when A is saturating.

What happens, however, if [A] is varied with [B] fixed but *non*-saturating?. In this case in equation (5.7) ϕ_0 and $\phi_B/[B]$ are both constant, while $\phi_A/[A]$ and $\phi_{AB}/[A][B]$ are both variable in [A], so that, regrouping,

$$\frac{e}{v} = \underbrace{\left[\phi_0 + \frac{\phi_B}{[B]}\right]}_{\text{constant term}} + \underbrace{\frac{1}{[A]}\left[\phi_A + \frac{\phi_{AB}}{[B]}\right]}_{\text{variable term}}. \qquad (5.10)$$

Both of the bracketed expressions in equation (5.10) are variable in 1/[B], but we have, of course, fixed [B]. Again, therefore, we have a linear equation in 1/[A], as when [B] was saturating, but now the actual values of the coefficients depend on [B]. At a given concentration of B the rate will vary with [A] with an *apparent* V_{max} given by

$$\frac{V_{max(app)}}{e} = \frac{1}{\phi_0 + \phi_B/[B]} \qquad (5.11)$$

and an *apparent* K_m given by

$$K_{A(app)} = (\phi_A + \phi_{AB}/[B])/(\phi_0 + \phi_B/[B]). \quad (5.12)$$

Analogous equations give the apparent V_{max} and K_m for B at different fixed concentrations of A.

We have already seen (5.9) that $K_{A(app)}$ at high [B] becomes equal to $\phi_A/\phi_0 = K_A$. Conversely, in equation (5.12), at very low [B], the second term in each bracket becomes dominant, so that $K_{A(app)}$ becomes equal to ϕ_{AB}/ϕ_B. Therefore, $K_{A(app)}$ can only be independent of [B] if $\phi_A/\phi_0 = \phi_{AB}/\phi_B$. Table 5.1 reveals that these expressions are only equal if

$$k_2 = \frac{k_5 k_7 k_9}{k_5 k_7 + k_5 k_9 + k_6 k_9 + k_7 k_9}$$

This arbitrary condition might by chance be approximately met in an individual case, but there are no grounds whatsoever for expecting it to be met.

Equation (5.12) tells us something of great practical importance. *The apparent* K_m *for one substrate depends on the concentration of the other substrate.* This is equally applicable to most other multi-substrate mechanisms. The measured. K_m's of multi-substrate enzymes are in no sense true constants, unless the fixed substrate(s) is/are saturating. The pitfall of the contrary assumption is inherent in the deceptively reassuring one-substrate analogy. Glutamate dehydrogenase (reaction 8, p. 37) provides a good illustration [44] of the scope for error. The K_m of this enzyme for NH_4^+, with fixed concentrations of the other substrates suitable for spectrophotometric assay (1 mM 2-oxoglutarate and 100 μM NADH), is approximately 40 mM. With the concentrations thought to obtain within the mitochondrion, however, (0.1 mM 2-oxoglutarate and 1 μM NADH) the K_m for NH_4^+ is 3–4 mM! A realistic appraisal of the enzyme's capacity for dealing with ammonia clearly depends on the use of appropriate kinetic constants. These comments apply equally to the use of V_{max} as an indication of the maximum possible flux through a reaction *in vivo.* The V_{max} measured under optimal conditions *in vitro* may well be unattainable in the living cell. For example, for enzymes utilizing NAD^+/NADH or ADP/ATP, an upper limit is imposed by the pool size of these nucleotides. The appropriate V_{max} would be an apparent one as defined by equation (5.11). Also it must be emphasized that these kinetic constants refer to initial-rate conditions; in the metabolic steady state, products as well as substrates are present, so that even the 'apparent V_{max}' is an overestimate. A realistic upper estimate may nevertheless be useful provided it is treated as such.

5.7 Dalziel ϕ parameters.

You may wonder why the ϕ symbols were introduced in equation (5.7). The equation can indeed be rearranged (equation 5.13) in a form [43] which bears a more obvious relationship to the Michaelis–Menten equation:–

$$v = \frac{V_{max}[A][B]}{K_{AB} + K_A[B] + K_B[A] + [A][B]} \quad (5.13)$$

The values of V_{max}, K_A, K_B and K_{AB} in terms of the rate constants of equation (5.6) for the compulsory-order ternary-complex mechanism are given in Table 5.1, and their relation to the ϕ parameters is shown in Table 5.2. The merit of the Michaelis–Menten equation, however, is that its parameters have an easily-grasped physical meaning. In equation (5.13) only V_{max} retains that meaning without qualification. As we have already seen, K_A and K_B are *not* the operational K_m's that metabolic biochemists might use. They are limiting values, approached when the fixed substrate is saturating. As for K_{AB}, although its mathematical significance is clear, it cannot be said to have an easily-grasped physical meaning.

Table 5.2 The relationships between the Dalziel and Alberty initial-rate parameters for a 2-substrate reaction

A. Dalziel parameters [42] in terms of Alberty parameters [43]

$\phi_0 = \frac{e}{V_{max}}$	$\phi_A = \frac{K_A e}{V_{max}}$	$\phi_B = \frac{K_B e}{V_{max}}$	$\phi_{AB} = \frac{K_{AB} e}{V_{max}}$

B. Alberty parameters in terms of Dalziel parameters

$V_{max} = \frac{e}{\phi_0}$	$K_A = \frac{\phi_A}{\phi_0}$	$K_B = \frac{\phi_B}{\phi_0}$	$K_{AB} = \frac{\phi_{AB}}{\phi_0}$

The ϕ parameters introduced by Dalziel [42] are more simply related to actual rate constants than the K's of equation (5.13)– e.g. Table 5.1. It is probably no coincidence that the important maximum-rate relationships (Section 5.10) were not discovered until the ϕ nomenclature was introduced [42]. Such relationships tend to be obscured in the algebraic profusion of the expressions for K_A, K_B and K_{AB}. For enzymes with more than two substrates the rate equations are still more cumbersome, reinforcing the argument for a nomenclature bearing the simplest possible relationship to the mathematical structure of the mechanism.

An attractive practical feature of the ϕ constants is the immediacy with which they emerge from the graphical analysis of kinetic results as discussed in the next section.

5.8 Two-substrate plotting procedure

If an enzyme really obeys equation (5.7/5.13), then, in theory, just four rate measurements with suitably chosen values of [A] and [B] should give simultaneous equations allowing one to solve for the four unknown constants. In practice, experimental measurements are not sufficiently accurate to allow such a procedure, and it would, in any case, involve pre-judging the issue by *assuming* equation (5.7) to be applicable.

Instead, one chooses a range of suitable concentrations, say six, of each substrate, and then measures the rate with all the possible combinations of these substrate concentrations, obtaining a good pair of duplicates for each set of conditions. One may then construct Lineweaver–Burk plots against either 1/[A] or 1/[B]; it does not matter which, and with good data, very similar estimates of the constants should be obtained by either route. In either case one obtains a plot for each fixed concentration of the second substrate (Fig. 5.1). If all these plots are linear, the next step is to measure their slopes and ordinate intercepts. The intercepts of primary plots against 1/[A] (Fig. 5.1) are given by $\phi_0 + \phi_B/[B]$ (see equation 5.7) and the slopes by $\phi_A + \phi_{AB}/[B]$. If these values cannot all be accurately read off the graphs, it may be necessary to measure a few more rates with strategically chosen substrate concentrations. One now plots the primary slopes (Fig. 5.2a) and intercepts (Fig. 5.2b) against 1/[B]. Taking the slopes first, the secondary plot will be linear, if

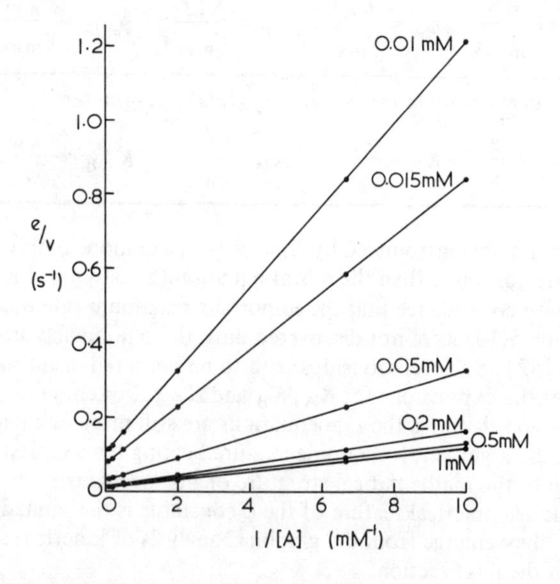

Fig. 5.1 Primary Lineweaver–Burk plots for a 2-substrate enzyme obeying equation (5.7). The lines are for six different fixed values of [B], as indicated. These are not real experimental results; if they were, some scatter would be expected. This plot would not allow accurate estimation of the slopes and intercepts of the shallower lines. In a real analysis these would be separately plotted on a larger scale.

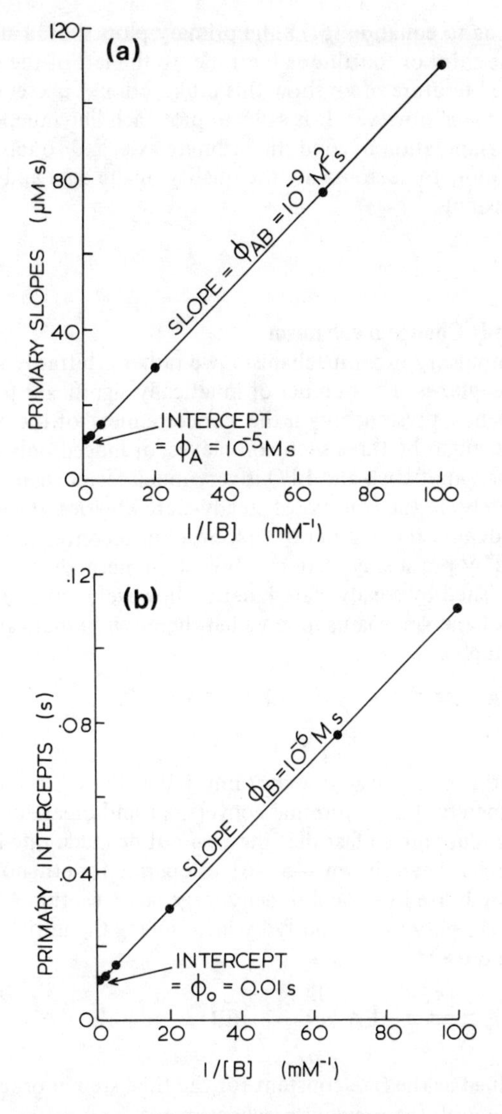

Fig. 5.2 Secondary replots. In (a) the slopes of the lines in Fig. 5.1 are plotted against the corresponding values of 1/[B]. In (b) the same is done for the ordinate intercepts from Fig. 5.1.

equation (5.7) applies, and will itself have a slope of ϕ_{AB} and an intercept of ϕ_A. Likewise, the secondary plot of intercepts should have a slope of ϕ_B and an intercept of ϕ_0. This simple series of plots thus gives the four constants of equation (5.7) directly.

According to equation (5.7), the primary plots should all intersect at a point in the third or fourth quadrant, i.e. to the left of the ordinate and plots in the literature often show this intersection. This tends to bias manual plotting, however. It is safer to plot each line independently, without extrapolation beyond the ordinate axis, and to rely on the secondary plots to demonstrate the quality of the data and provide the kinetic constants.

5.9 Theorell–Chance mechanism

In our compulsory-order mechanism, we rather arbitrarily included two ternary complexes. The number of kinetically significant ternary complexes depends, presumably on the 'concertedness' of the catalytic conversion. There might be three such complexes, or indeed only one, if the interconversion of EAB and EPQ is very much faster than all other steps. Unfortunately, in the context of steady-state kinetics, the argument is purely academic: the number of first-order interconversions can only be established by pre-steady-state rapid-reaction methods. One case that can be distinguished by steady-state kinetics, however, is the so-called Theorell–Chance mechanism, in which there is *no* kinetically significant ternary complex.

$$E + A \rightleftharpoons EA \underset{B \quad Q}{\overset{}{\rightleftharpoons}} EP \rightleftharpoons E + P.$$

Note that this mechanism does *not* imply that there *is* no ternary complex, but merely that its internal conversions and breakdown to release the first product are so fast that they cannot be made rate-limiting. This mechanism has been shown [45, 46] to operate for ethanol oxidation catalysed by horse liver alcohol dehydrogenase (reaction 4, p. 22).

We can simplify as previously by introducing the initial-rate assumption at the outset:–

$$E + A \underset{k_2}{\overset{k_1}{\rightleftharpoons}} EA \xrightarrow[Q]{B \; k_3} EP \xrightarrow{k_9} E.$$

(k_9 is retained as the rate constant for the final step in order to facilitate comparison with the previous mechanism.)

Steady state for EP: $[EP]\,k_9 = [EA]\;k_3\;[B]$

$$\therefore [EA] = \frac{[EP]\,k_9}{k_3\,[B]} \tag{5.14}$$

Steady state for EA: $[EA]\;(k_3\,[B] + k_2) = [E]\,k_1\,[A]$

$$\therefore \quad [E] = [EA]\,\frac{(k_3\,[B] + k_2)}{k_1\,[A]} = [EP]\,\frac{k_9(k_3\,[B] + k_2)}{k_1 k_3\,[A]\,[B]} \tag{5.15}$$

Enzyme conservation: $e = [EP] + [EA] + [E]$

$$= [EP]\left[1 + \frac{k_9}{k_3[B]} + \frac{k_9(k_3[B] + k_2)}{k_1 k_3 [A][B]}\right] \quad (5.16)$$

$$v = k_9[EP] \therefore \frac{e}{v} = \frac{1}{k_9} + \frac{1}{k_3[B]} + \frac{1}{k_1[A]} + \frac{k_2}{k_1 k_3 [A][B]}. \quad (5.17)$$

This again is of the same form as equation (5.7), with the values of the coefficients as listed in Table 5.3.

Table 5.3 The kinetic parameters for a Theorell-Chance mechanism

ϕ_0	ϕ_A	ϕ_B	ϕ_{AB}	ϕ_0'	ϕ_P	ϕ_Q	ϕ_{PQ}
$\frac{1}{k_9}$	$\frac{1}{k_1}$	$\frac{1}{k_3}$	$\frac{k_2}{k_1 k_3}$	$\frac{1}{k_2}$	$\frac{1}{k_{10}}$	$\frac{1}{k_8}$	$\frac{k_9}{k_8 k_{10}}$

The first four parameters are for the forward reaction (equation 5.17). The other four are the corresponding and symmetrical parameters for the reverse reaction.

5.10 Tests of mechanism

Earlier we stated that the kinetic parameters obtained for an enzyme may be tested against the predictions for various mechanisms. We now have examples to work with, and may pause to consider the nature of kinetic tests before deriving any more rate equations. The tests are of three basic types [42]:

(i) presence or absence of individual terms in the rate equation. If the kinetic pattern is non-linear, i.e. if equation (5.7) is not an adequate description, there may be terms in $[A]^2$ or $[B]^2$ etc. If the pattern is linear, the rate equation may contain all four terms of equation (5.7), but, on the other hand, individual terms may be missing;

(ii) comparison of kinetic parameters or their ratios. Various internal equalities or inequalities may be predicted for individual mechanisms. Some involve only the kinetic parameters for reaction in one direction. Other comparisons involve parameters for both forward and reverse reactions;

(iii) it is also possible to compare the kinetic parameters or their ratios with such independently measurable quantities as the overall equilibrium constant of the catalysed reaction, dissociation constants for binding of substrates to free enzyme, and individual rate constants which may be measurable by rapid-reaction techniques.

Let us now apply these tests to the two mechanisms we have considered. We can convert the first mechanism into a Theorell–Chance mechanism by making certain rate constants very large relative to the rest; setting k_5 and k_7 very large converts equation (5.6) into equation (5.17). One may therefore expect that the Theorell–Chance mechanism, the special case, will share the algebraic characteristics of the more general mechanism, but that it will also have distinguishing features of its own.

If we first consider the form of the equation, the mechanisms are indistinguishable. Both give a linear equation (5.7) containing all four terms (5.6 and 5.17). Looking more closely at the expression for the ϕ constants (Tables 5.1 and 5.3), we discover four useful relationships common to both mechanisms:–

$$\phi_{AB}/\phi_B = k_2/k_1 \tag{5.18}$$

$$\phi_A = 1/k_1 \tag{5.19}$$

$$\frac{\phi_{AB}}{\phi_A \phi_B} = k_2 \tag{5.20}$$

$$\frac{\phi_{AB}}{\phi_{PQ}} = K_{eq} \tag{5.21}$$

The ratio k_2/k_1 in equation (5.18) is the dissociation constant for A. This may be measured directly in the absence of B, by equilibrium dialysis for example, and compared with the kinetic estimate. If the two values are not in reasonable agreement, one can rule out all mechanisms that predict the equality. If the enzyme will work with several alternative second substrates, B_1, B_2 etc., one may determine the kinetic parameters with each one, and compare the values of ϕ_{AB}/ϕ_B. Since k_2 and k_1 do not involve substrate B in any way, ϕ_{AB}/ϕ_B should not change when B_2 is substituted for B_1. An example is the use of different alcohols with horse liver alcohol dehydrogenase [47]. It is important to remember that a compulsory-order mechanism may have been formulated the wrong way round: if B is the leading substrate, then ϕ_{AB}/ϕ_B is not independent of the nature of B; instead ϕ_{AB}/ϕ_A is independent of the nature of A.

Similarly, equation (5.19) predicts the value of a single rate constant, k_1, which may be directly measurable and should again be independent of the nature of B if A leads.

Equation (5.20) follows directly if (5.18) and (5.19) are satisfied, but forms the basis of another test. The first-order constant k_2 governs a compulsory product-release step in the *reverse* reaction. This means that under no circumstances can the maximum rate of the reverse reaction exceed $k_2 e$, provided that the compulsory-order mechanism is valid. If the initial-rate parameters are determined for both directions of reaction this relationship may be tested:–

$$\frac{\phi_{AB}}{\phi_A\phi_B} = k_2 \geqslant \frac{1}{\phi_0'}$$

$$\frac{\phi_{PQ}}{\phi_P\phi_Q} = k_9 \geqslant \frac{1}{\phi_0}.$$

These maximum rate tests, known as the Dalziel relationships, help not only to distinguish compulsory-order sequential mechanisms from others, but also to distinguish the Theorell–Chance mechanism from those in which there is a kinetically significant ternary complex. For the more general mechanism k_2 defines an upper limit for $1/\phi_0'$. For the Theorell–Chance mechanism, however, k_2 *equals* $1/\phi_0'$ (Table 5.3). This is evident if one examines the mechanism: saturating with P and Q forces all the enzyme into the EA complex, so that the rate is entirely limited by the breakdown of this complex, governed by k_2. If, therefore, k_2 is significantly greater than $1/\phi_0'$, the Theorell–Chance mechanism may be ruled out. The fact that $1/\phi_0' = k_2$, and that $1/\phi_0 = k_9$, for a Theorell–Chance mechanism, may be used in another way as a test of the mechanism if the enzyme shows a broad specificity for its 'inner' substrate. In the mechanism on p. 56, k_9 and k_2 are independent of the nature of B and Q. Thus, if it is possible to use several different substrates, the maximum rate should be the same for all of them. This is the case for liver alcohol dehydrogenase with different primary alcohols, NAD^+ and NADH being the 'outer' substrates [47].

Finally we have the Haldane relationship between the kinetic parameters and the overall equilibrium constant (5.21). The use of Haldane relationships requires initial-rate analysis of reaction in both directions. For the more general of the two mechanisms we have considered,

$$K_{eq} = \frac{k_2k_4k_6k_8k_{10}}{k_1k_3k_5k_7k_9}.$$

Equation (5.21) for this mechanism may be verified by substituting the value of ϕ_{AB} from Table 5.1, and writing down from considerations of symmetry the corresponding value of ϕ_{PQ}:–

$$\phi_{PQ} = \frac{k_9(k_4k_6 + k_4k_7 + k_5k_7)}{k_4k_6k_8k_{10}}.$$

For the Theorell–Chance mechanism there are only six rate constants, and

$$K_{eq} = \frac{k_2k_8k_{10}}{k_1k_3k_9}.$$

The constants in Table 5.3 again readily confirm equation (5.21), but they also give another distinctive Haldane relationship:–

$$K_{eq} = \frac{\phi_0' \phi_P \phi_Q}{\phi_0 \phi_A \phi_B}. \tag{5.22}$$

5.11 Random-order mechanisms

Enzymes must display some degree of specificity for all their substrates, so that in some senses a random order of substrate addition might seem the most likely course of events in a ternary-complex mechanism. Kineticists tend to favour compulsory-order mechanisms, but this may be wishful thinking. Steady-state random-order mechanisms generate much more formidable rate equations. By definition, in such mechanisms each substrate can combine with more than one enzyme form, and these multiple points of combination are connected through reversible steps. If substrate A can bind not only to the free enzyme, E, but also to the binary complex, EB, then the concentration of E available to form EA is affected by the binding of A to EB. Conversely the concentration of EB is influenced by the depletion of E to form EA. As a result, the expressions for the concentrations of the various enzyme species contain terms in $[A]^2$ as well as $[A]$. This can readily be verified by partial steady-state analysis of a suitable piece of mechanism. Squared terms lead to non-linear Lineweaver–Burk plots and non-hyperbolic plots of initial rate against substrate concentration. Such plots are often obtained experimentally, and are discussed further in Chapter 7.

It is also possible, however, for a random-order mechanism to give linear kinetics, and several examples are known (e.g. [48]). To obtain a linear equation we have to abandon the steady-state assumption, and return to the original Michaelis–Menten postulate of a rate-limiting catalytic step, with all preceding steps at equilibrium. This is then known as a *rapid-equilibrium random-order* mechanism:–

```
     KA   EA   KB(A)
   ↙↗          ↘↖
 E                EAB ——k——→ E + PRODUCTS
   ↘↖          ↙↗
     KB   EB   KA(B)
```

Since, in this mechanism, the overall rate is $k[EAB]$, it is simplest to express the other concentrations in terms of $[EAB]$. The concentrations are all related by dissociation constants, rather than rate constants, as follows:–

$$K_{B(A)} = \frac{[EA][B]}{[EAB]} \qquad \therefore \quad [EA] = [EAB]\frac{K_{B(A)}}{[B]}$$

$$K_A = \frac{[E][A]}{[EA]} \qquad \therefore \quad [E] = [EA]\frac{K_A}{[A]} = [EAB]\frac{K_{B(A)}K_A}{[B][A]}$$

$$K_B = \frac{[E][B]}{[EB]} \qquad \therefore \quad [EB] = [E]\frac{[B]}{K_B} = [EAB]\frac{K_{B(A)}K_A}{K_B[A]}$$

Enzyme conservation: $e = [EAB] + [EB] + [EA] + [E]$

$$= [EAB]\left[1 + \frac{K_{B(A)}K_A}{K_B[A]} + \frac{K_{B(A)}}{[B]} + \frac{K_{B(A)}K_A}{[A][B]}\right]$$

∴ Since $v = k[EAB]$,

$$\frac{e}{v} = \frac{1}{k}\left[1 + \frac{K_{B(A)}K_A}{K_B[A]} + \frac{K_{B(A)}}{[B]} + \frac{K_{B(A)}K_A}{[A][B]}\right] \qquad (5.23)$$

Since $K_A K_{B(A)} = K_B K_{A(B)}$, the second term in equation 5.23 can also be written as $K_A/[A]$.

Again the equation is of the same form as equation (5.7), predicting linear Lineweaver–Burk plots intersecting in the third or fourth quadrant. The algebraic expressions for the ϕ parameters are given in Table 5.4. As shown in the same Table, for this mechanism all the dissociation constants may be obtained as ratios of paired ϕ constants. Those involving the free enzyme, K_A and K_B, are valuable as tests of the mechanism since they may be independently measured. The mere demonstration that both substrates can bind to the free enzyme points towards a random-order mechanism, but quantitative agreement between directly measured dissociation constants and kinetic estimates provides more solid evidence.

Note that, of the expressions for K_A and K_B in Table 5.4, the first would be true of a compulsory-order mechanism with A binding first, and the second would be true of a mechanism with B leading, but only for the rapid-equilibrium random-order mechanism are both expressions true. The validity of these relationships can be further tested if data are obtainable for a series of related substrates. Ideally substitutes for both A and B should be used, but the specificity of an enzyme does not always allow this.

By setting up the equation for the reverse reaction it may easily be shown that the Haldane relationship for this mechanism is still given by equation (5.21), as for the two other sequential mechanisms previously considered.

Table 5.4 Rapid-equilibrium random-order mechanism: relationships between the ϕ parameters and the dissociation constants

$\phi_0 = \dfrac{1}{k}$	$\phi_A = \dfrac{K_{B(A)}K_A}{K_B k} = \dfrac{K_{A(B)}}{k}$
$\phi_B = \dfrac{K_{B(A)}}{k}$	$\phi_{AB} = \dfrac{K_{B(A)}K_A}{k}$
$K_A = \phi_{AB}/\phi_B$	$K_B = \phi_{AB}/\phi_A$
$K_{A(B)} = \phi_A/\phi_0$	$K_{B(A)} = \phi_B/\phi_0$

The K_m for each substrate in this mechanism (i.e. at infinite concentration of the other substrate) is equal to the dissociation constant for dissociation of the substrate *from the ternary complex*. It is only equal to the constant for dissociation from the binary complex in the special case where A and B do not affect one another's binding, i.e. when $K_A = K_{A(B)}$ and $K_B = K_{B(A)}$.

5.12 Enzyme substitution or ping-pong mechanism

The ping-pong mechanism stands apart from the others. In this mechanism A and B are never on the enzyme surface together. The substrate binding steps are separated by product release steps, by definition irreversible under initial-rate conditions since product is at zero concentration

$$\mathrm{E} \underset{k_2}{\overset{k_1[\mathrm{A}]}{\rightleftharpoons}} \mathrm{EA} \xrightarrow{k_3} \mathrm{P} + \mathrm{E'} \underset{k_6}{\overset{k_5[\mathrm{B}]}{\rightleftharpoons}} \mathrm{E'B} \xrightarrow{k_7} \mathrm{Q} + \mathrm{E}$$

Accordingly, in the scheme above, the k_4 and k_8 (product addition) steps are omitted.

In the normal way, the steady state for E′B gives:–

$$[\mathrm{E'}]\, k_5\, [\mathrm{B}] = [\mathrm{E'B}]\, (k_6 + k_7). \tag{5.24}$$

For the next step we can take a short cut. In the steady state, the net flow at each step, from E to EA, from EA to E′, and so on, must be the same. Hence immediately:–

$$[\mathrm{EA}]\, k_3 = [\mathrm{E'B}]\, k_7. \tag{5.25}$$

Finally, the steady state for EA gives:

$$[\mathrm{E}]\, k_1\, [\mathrm{A}] = [\mathrm{EA}]\, (k_2 + k_3).$$

∴ substituting from equation (5.25),

$$[\mathrm{E}] = [\mathrm{E'B}]\, \frac{k_7(k_2 + k_3)}{k_1 k_3 [\mathrm{A}]}. \tag{5.26}$$

Enzyme conservation:

$$e = [\mathrm{E'B}] + [\mathrm{E'}] + [\mathrm{EA}] + [\mathrm{E}]$$

$$= [\mathrm{E'B}]\left[1 + \frac{(k_6 + k_7)}{k_5[\mathrm{B}]} + \frac{k_7}{k_3} + \frac{k_7(k_2 + k_3)}{k_1 k_3 [\mathrm{A}]}\right] \tag{5.27}$$

$$v = k_7\, [\mathrm{E'B}]$$

$$\therefore\ \frac{e}{v} = \frac{1}{k_7} + \frac{1}{k_3} + \frac{(k_2 + k_3)}{k_1 k_3 [\mathrm{A}]} + \frac{(k_6 + k_7)}{k_5 k_7 [\mathrm{B}]}. \tag{5.28}$$

The ϕ constants are listed in Table 5.5.

The striking thing about equation (5.28) is that one of the four terms of equation (5.7) is missing [42]. There is no term in $1/[\mathrm{A}][\mathrm{B}]$. This is an inevitable consequence of the pattern of substrate addition and

Table 5.5 Kinetic parameters for a two-substrate ping-pong mechanism

$\phi_0 = \dfrac{1}{k_7} + \dfrac{1}{k_3} = \dfrac{k_3 + k_7}{k_3 k_7}$	$\phi_A = \dfrac{k_2 + k_3}{k_1 k_3}$	$\phi_B = \dfrac{k_6 + k_7}{k_5 k_7}$
$\phi_0' = \dfrac{1}{k_2} + \dfrac{1}{k_6} = \dfrac{k_2 + k_6}{k_2 k_6}$	$\phi_Q = \dfrac{k_6 + k_7}{k_6 k_8}$	$\phi_P = \dfrac{k_2 + k_3}{k_2 k_4}$

product release. The distribution of enzyme between E and EA depends on [A], and the distribution between E′ and E′B depends on [B], but the distribution between E plus EA, on the one hand, and E′ plus E′B, on the other, depends only on the relative values of k_3 and k_7 (cf. equation (5.24)). This is because these two halves of the mechanism are separated by irreversible steps.

How does the absence of the ϕ_{AB} term affect the plots? A plot of e/v against $1/[A]$ has an ordinate intercept of $\phi_0 + \phi_B/[B]$. The slope, however, is simply ϕ_A. This is independent of [B], and so the Lineweaver-Burk plots are a series of parallel lines (Fig. 5.3). This is of course also true for the plots of e/v against $1/[B]$. A secondary replot of the intercepts as in Fig. 5.2b is necessary in order to evaluate ϕ_0, but the replot of slopes is superfluous: if the primary slopes are really constant the replot must be a horizontal line.

The parallel-line pattern in the Lineweaver–Burk plots for a ping-pong mechanism is the most striking and immediate kinetic criterion of mechanism that we have so far encountered. It is reinforced by a distinctive Haldane relationship. Clearly equation (5.21) cannot apply, since ϕ_{AB} and ϕ_{PQ} are absent from the rate equation. As usual we can use the symmetry of the mechanism to write down the kinetic parameters for the reverse reaction by inspection (Table 5.5). Examination of the constants reveals that:

$$\frac{\phi_A \phi_B}{\phi_P \phi_Q} = \frac{k_2 k_4 k_6 k_8}{k_1 k_3 k_5 k_7} = K_{eq} \qquad (5.29)$$

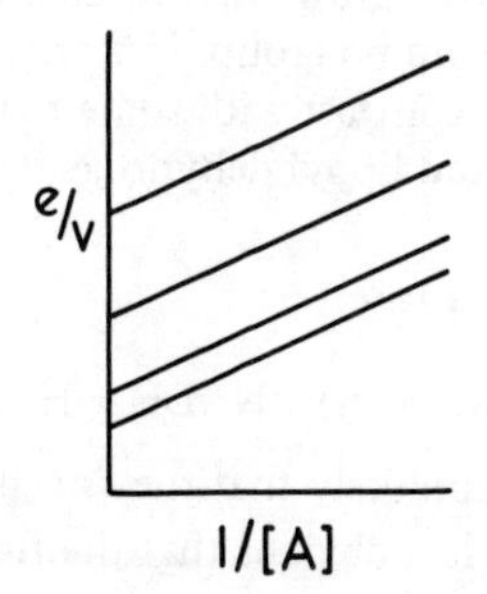

Fig. 5.3 Ping-pong kinetics. The characteristic pattern of parallel Lineweaver–Burk plots given by enzymes that catalyse a 2-substrate reaction not involving a ternary complex.

The observance of this relationship, taken together with parallel Lineweaver–Burk plots, would constitute strong evidence for a ping-pong mechanism. It is not always possible to study a reaction in both directions, however. If one is forced to rely on the evidence of parallel plots alone, it is important to be certain that they are truly parallel. This is less easy than it might seem. A ϕ_{AB} term may be present but be masked under the conditions of experimentation. From equation (5.10) (p. 51), it can be seen that, as [B] is raised, the slopes of plots of e/v against $1/[A]$ gradually become constant even if there is a ϕ_{AB} term. Constancy results when $[B] \gg \phi_{AB}/\phi_A$. This is part of the process of saturation with B, and at the same time the contribution of $\phi_B/[B]$ becomes less significant in the expression for the intercept (equation 5.10). There are no grounds, however, for expecting the ϕ_{AB} and ϕ_B terms to 'disappear' over the same range of [B]. The intercept becomes constant when $[B] \gg \phi_B/\phi_0$. Thus, if $\phi_B/\phi_0 \gg \phi_{AB}/\phi_A$, then, for values of [B] lying between these two limits, Lineweaver–Burk plots against $1/[A]$ will have variable intercepts but constant slope – i.e. they will appear parallel. The same inequality predicts parallel plots against $1/[B]$ for values of [A] lying between ϕ_A/ϕ_0 and ϕ_{AB}/ϕ_B.

Under the conditions just considered, the ϕ_{AB} term is dwarfed by both the ϕ_A and ϕ_B terms. In fact, even if only one of these terms is much larger than the ϕ_{AB} term, the latter may be difficult to detect, partly because of the deception of the eye, and partly because the rate variation due to the ϕ_{AB} term may not be much greater than the inherent uncertainty of the measurements. Let us suppose the ϕ_A term to be dominant. In the plots against $1/[A]$ the lines will appear steep and very nearly parallel (Fig. 5.4a). In plots against $1/[B]$ the slopes will vary more in percentage terms, because the ϕ_B term is not dominant, but this may still not be evident, because all the slopes will be shallow, and all the lines will be widely spaced out by the ϕ_A term (Fig. 5.4b).

One remedy is to use lower ranges of substrate concentration, but limits are always imposed by the sensitivity of the measuring technique. The lengths to which one pursues the possibility that apparently parallel lines may not be truly parallel depends on the individual case. The chemical symmetry of reaction (2) (p. 11), for example, and the knowledge that a prosthetic group is involved, point towards a mechanism in which Product No. 1 has to vacate the active site so that its analogue, Substrate No. 2, may come in to receive the amino group. In a sense the kinetic evidence confirms intuition here. By contrast with some flavoproteins, e.g. glucose oxidase (reaction 10) and lipoyl dehydrogenase (reaction 11),

$$\text{D-Glucose} + O_2 \longrightarrow \text{D-gluconate} + H_2O_2 \qquad \text{... (10)}$$

$$\text{Lipoate(SH)}_2 + NAD^+ \rightleftharpoons \text{Lipoate(S–S)} + NADH + H^+ \qquad \text{... (11)}$$

the substrates are so dissimilar that it seems unlikely that they occupy the same site on the enzyme. It is therefore less obvious that the first

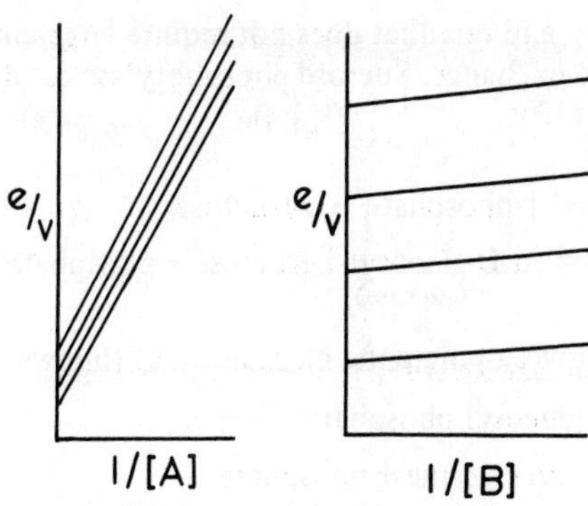

Fig. 5.4 Lineweaver–Burk plots for an enzyme-catalysed reaction under conditions in which the ϕ_{AB} term is dwarfed by the much larger contribution of the ϕ_A term. In both plots the lines actually converge slightly, but this is difficult to see. The slight slope variation would be beyond the precision of most kinetic measurements and would therefore probably remain undetected.

product must leave before the second substrate may bind. Indeed, although reaction 11 has been thought to follow ping-pong kinetics [49,50], more recent evidence suggests that it may after all involve a ternary complex [51].

Koster and Veeger [52] have shown that competitive inhibitors can be useful in revealing a hidden ϕ_{AB} term. Thus, for D-amino acid oxidase, they obtained plots in the absence of inhibitor similar to those in Fig. 5.4. By adding benzoate, which is competitive with the amino acid substrate, they enhanced the contribution of the two terms in 1/[amino acid], so that the ϕ_{AB} term became manifest.

As with other mechanisms, interchangeable substrates help in testing a ping-pong mechanism. Comparison of the ϕ parameters in Table 5.5 with the mechanism itself shows that ϕ_A should be unchanged when a different substrate B is used, and likewise ϕ_B should be independent of the nature of A. Dixon and Kleppe [53] adduced the constancy of ϕ_{oxygen} with a range of amino acid substrates as evidence for a ping-pong mechanism for D-amino acid oxidase. (As we have seen above, that mechanism has since been challenged.)

For a ping-pong mechanism it should be possible to demonstrate 'partial reactions'. Thus, in the mechanism illustrated on p. 62, conversion of A to P should occur in the absence of B. Net conversion in such a partial reaction cannot, however, exceed the total amount (in molar terms) of enzyme present, since the enzyme is itself one of the reactants. In order to measure a net conversion, therefore, a large amount of enzyme may be needed. If this is possible, it may also be feasible to detect the change in the enzyme itself. Thus, the reduction of a flavoprotein may readily be detected spectrophotometrically, although this does not necessarily imply that the product has been released. The neatest method of demonstrating a partial reaction, and one that does not require large amounts of enzyme, is that of isotope exchange. Sucrose phosphorylase catalyses, amongst others, reaction (12):

$$\alpha\text{-D-glucosyl-1-phosphate} + \text{D-fructose} \rightleftharpoons \underset{\text{(sucrose)}}{\alpha\text{-D-glucosyl-1-fructose}} + \text{phosphate} \quad \ldots (12)$$

This enzyme follows a ping-pong mechanism as follows:

enzyme + glucosyl phosphate $\longleftrightarrow$
glucosyl enzyme + phosphate

glucosyl enzyme + fructose $\longleftrightarrow$
glucosyl fructose + enzyme

The evidence for these steps comes from experiments in which:

(a) the enzyme catalysed rapid incorporation of label from radioactive inorganic phosphate into α-D-glucosyl-1-phosphate in the absence of fructose [54];

(b) in the absence of phosphate, label from radioactive fructose was incorporated into initially unlabelled sucrose.

5.13 Three substrates

Reactions with more than two substrates are not uncommon. Ligase reactions involve at least three substrates, and many dehydrogenase reactions also involve three substrates in one direction. Few such reactions have been thoroughly kinetically investigated. It is generally assumed that the analysis must be intolerably complex. Certainly an extra substrate greatly magnifies the experimental task of exploring sufficient combinations of substrate concentrations, but the analysis is in some ways easier than for 2-substrate reactions.

For a two-substrate enzyme obeying linear kinetics equation (5.7) gives the maximum possible number of terms in the reciprocal form of the initial-rate equation, and, as we have seen, for any mechanism with a ternary complex, all four terms are present. For a 3-substrate enzyme that gives linear Lineweaver–Burk plots, the corresponding equation contains 8 terms (equation 5.30):

$$\frac{e}{v_0} = \phi_0 + \frac{\phi_A}{[A]} + \frac{\phi_B}{[B]} + \frac{\phi_C}{[C]} + \frac{\phi_{AB}}{[A][B]} + \frac{\phi_{AC}}{[A][C]} + \frac{\phi_{BC}}{[B][C]} + \frac{\phi_{ABC}}{[A][B][C]} \quad (5.30)$$

In this case, however, a mechanism with a central quaternary complex EABC does not necessarily give rise to an equation with all 8 terms present. For a compulsory-order 3-substrate mechanism, the term in the first and third substrate, i.e. the term in [A] [C] if the order is A, B, C, is always missing, as Frieden first showed [55]. Thus *from the form of the equation alone* one can distinguish compulsory-order from rapid-equilibrium random-order, and, moreover, assign the second substrate if the mechanism is compulsory-order

This theory has been successfully applied to bovine liver glutamate dehydrogenase (reaction 8, p. 37). Two possible compulsory-order mechanisms had been suggested, both with coenzyme binding first, so that either the [NADH] [2-oxoglutarate] or [NADH] [NH_4^+] terms should have been missing. In fact, a systematic study involving primary, secondary and tertiary plotting of the data [44], led to an equation containing all the 8 terms. The process was repeated with two coenzymes and at two pH's. The results suggested a rapid-equilibrium random-order mechanism, since kinetic estimates of dissociation constants for 2-oxoglutarate and NH_4^+ obtained from the kinetic parameters on the assumption of such a mechanism remained constant on changing the coenzyme. This mechanism has since been supported by isotope exchange experiments [55a], and by the demonstration of some of the complexes that would be forbidden in a compulsory-order mechanism e.g. E-2-oxoglutarate.

The tertiary plotting procedure has also been applied to glyceraldehyde 3-phosphate dehydrogenase from pig muscle [56], and more recently to pyruvate kinase from yeast [57], showing that the catalysed reaction is reaction (13):

$$\text{Phosphoenolpyruvate} + Mg^{++} + \text{ADP} \rightleftharpoons \text{pyruvate} + (Mg^{++}\text{—ATP}) \quad \text{... (13)}$$

Previously it has been assumed on the basis of N.M.R. experiments that the nucleotide always reacted as the metal chelate, in both directions of reaction. The formulation above leads to a much more satisfying account of events at the active site, with the Mg^{++} acting as the bridge between the β-phosphate of ADP and the incoming phosphoryl group destined to be the γ-phosphate of ATP.

The mechanisms considered above for glutamate dehydrogenase are the two extremes of a range of possibilities. One can also visualize partially-random mechanisms for a three-substrate enzyme. Attempts have been made to derive linear equations for such mechanisms by rapid-equilibrium treatment [58], but these equations cannot be strictly valid over wide ranges of substrate concentration [59]. Consider, for example, the mechanism below, in which A must add first but B and C can then add randomly:

$$\text{E} \underset{k_2}{\overset{k_1}{\rightleftharpoons}} \text{EA} \begin{matrix} \nearrow\!\!\swarrow \ \text{EAB} \ \searrow\!\!\nwarrow \\ \searrow\!\!\nwarrow \ \text{EAC} \ \nearrow\!\!\swarrow \end{matrix} \text{EABC} \xrightarrow{k} \text{PRODUCTS}$$

A glance back at the way in which the rate equation for the 2-substrate rapid-equilibrium random-order mechanism arises (p. 60) allows us to tell by inspection that the rate equation for the mechanism above contains only five terms, a ϕ_0 term representing [EABC], ϕ_C and ϕ_B terms for [EAB] and [EAC] respectively, a ϕ_{BC} term for [EA], and a ϕ_{ABC} term representing the contribution of E.

$$\frac{e}{v_0} = \phi_0 + \frac{\phi_B}{[B]} + \frac{\phi_C}{[C]} + \frac{\phi_{BC}}{[B][C]} + \frac{\phi_{ABC}}{[A][B][C]} . \tag{5.31}$$

There is no ϕ_A term in equation (5.31) because there is no EBC complex. The absence of ϕ_A implies that when [B] and [C] are saturating, the rate is independent of the concentration of A. This is evidently absurd; it must be possible to make addition of A rate-limiting.

In fact, for any mechanism containing a compulsory segment, it is impossible to maintain rapid equilibrium among the complexes for all combinations of substrate concentration. If the five complexes in the mechanism above are to remain in equilibrium, the catalytic step must be slower than any compulsory step in the mechanism.

$$\therefore \; k[EABC] \ll k_1 \, [E]\,[A] .$$

If EABC is in equilibrium with E, this may be rewritten:

$$K'[E]\,[A]\,[B]\,[C] \ll k_1 \, [E]\,[A]$$

where K' is a constant incorporating k and the relevant equilibrium constants.

$$\therefore \; K'[B]\,[C] \ll k_1 .$$

This condition can clearly be violated by raising the concentration of B or C sufficiently.

By contrast, in the fully random mechanism there is no compulsory first step. If [A] is very small, and [B] and [C] are large relative to the dissociation constants for these substrates, the reaction proceeds almost entirely via EB, EC and EBC. Eventually A has to add (with a rate constant k_A), and so the condition for continuing equilibrium is that

$$k[EABC] \ll k_A [EBC]\,[A]$$

$$\therefore \; K'[E]\,[A]\,[B]\,[C] \ll K''[E]\,[A]\,[B]\,[C] .$$

Unlike the earlier inequality, this one depends only on the relative values of the constants: if it is satisfied for one set of values of [A], [B], and [C] it will be satisfied for all values of [A], [B] and [C].

In addition to the various quaternary complex mechanisms considered above, various enzyme substitution mechanisms are possible for a three-substrate reaction, all leading to deletion of terms in equation (5.30). These are comprehensively listed by Fromm [60].

5.14 Inhibition of multi-substrate reactions

The inhibition of enzymes with more than one substrate is of interest for all the reasons enumerated in Chapter 3, but also as a means of distinguishing alternative mechanisms.

Reversible inhibitors which act by diverting enzyme from the reaction pathway, as considered in Chapter 3, may be termed 'dead-end' inhibitors. Such inhibitors are not substrates for the enzyme, although they may be substrate analogues. For multi-substrate enzymes there is also the possibility of 'product inhibition'. A single product without its reaction partners cannot initiate the overall reverse reaction, but it can nevertheless decrease the initial rate of the forward reaction by opposing one of its steps. Thus a product inhibitor does not divert enzyme from the normal reaction pathway, but it alters the enzyme's distribution among the complexes along that pathway.

Products also sometimes form 'abortive complexes'. For example, pyruvate as a substrate of lactate dehydrogenase (reaction 9, p. 48) binds to the E.NADH complex. As a product inhibitor of lactate oxidation, therefore, it slows down reaction by opposing the dissociation of pyruvate from the ternary complex. Also, however, pyruvate forms the abortive complex E • pyruvate • NAD^+ [61], which is not part of the main reaction pathway. Thus pyruvate acts as a dead-end inhibitor as well as a simple product inhibitor.

Despite these complexities, the inhibition is still classified as competitive, uncompetitive, or non-competitive, depending on whether it affects the slope of the Lineweaver–Burk plot, the intercept, or both. The varied substrate must be specified, however, because an inhibitor non-competitive with respect to substrate A may be competitive with respect to substrate B, and so on. The pattern of inhibition is established with respect to each of the substrates in any thorough study.

As in the one-substrate case, K_i's may be evaluated for multi-substrate enzymes. It is often assumed that a K_i is automatically equal to the dissociation constant of the enzyme-inhibitor complex El. In fact:–

(i) the inhibitor, as we have seen, may combine with more than one enzyme complex, often with different dissociation constants;

(ii) if both the slope and the intercept of the Lineweaver–Burk plot are affected, there may be different K_i's for the two effects;

(iii) even if the inhibitor combines at only one point in the mechanism, the apparent slope or intercept K_i, determined as described in Chapter 3, usually depends on the concentration(s) of the fixed substrate(s). As shown earlier, the primary slope or intercept may be the sum of two or more terms. If only one of these is affected by the inhibition, or if the terms are affected to different extents, the apparent overall K_i for the slope or intercept will not be the same as the true K_i for the individual term(s). Thus, if

$$\text{UNINHIBITED SLOPE} = \text{TERM 1} + \text{TERM 2}$$

and the inhibition only affects TERM 1, then

$$\text{INHIBITED SLOPE} = \text{TERM 1}\left(1 + \frac{[\text{I}]}{K_i}\right) + \text{TERM 2}$$

If, therefore, the slope is plotted against [I] (Chapter 3), the abscissa intercept, giving the apparent K_i will be

$$K_i[\text{TERM 1} + \text{TERM 2}]/\text{TERM 1}.$$

Clearly, to predict the initial rate for all combinations of substrate and inhibitor concentration, one needs the true K_i for each term affected by the inhibitor.

(iv) even when the K_i for an individual term in the rate equation is known, its relation to dissociation constants depends on the details of the mechanism. A single term may represent more than one enzyme complex – e.g. in the 2-substrate ping-pong mechanism (p. 62) ϕ_0 represents both E′B and EA.

How do we set about predicting the inhibition patterns for a multi-substrate mechanism? Consider the mechanism below:–

$$E \rightleftharpoons EA \rightleftharpoons EAB \rightleftharpoons EPQ \longrightarrow EP \longrightarrow E$$

From the steady-state treatment (p. 49) we know that the ϕ_0 term is contributed by EP, EPQ and EAB, the ϕ_B term by EA, and both the ϕ_A and ϕ_{AB} terms by E. This immediately tells us that a dead-end inhibitor, I_1 combining, say, with EA, increases ϕ_B and thus is competitive with respect to B (only slope changes) and uncompetitive with respect to A (only intercept changes). Likewise, a dead-end inhibitor, I_2, combining with E, increases both the ϕ_{AB} and ϕ_A terms, and thus, whilst it is competitive with respect to A, it is *non*-competitive with respect to B (both slope and intercept change).

Cleland [62] has, however, provided rules for predicting dead-end or product inhibition patterns by inspection of the mechanism without working out the rate equation. These useful rules are explained below.

An intercept effect means that saturation with the variable substrate does not abolish the inhibition. Accordingly intercepts are altered when the inhibitor combines with an enzyme form other than that with which the variable substrate combines – e.g. I_1 with A as the variable substrate above.

A slope effect is obtained if a dead-end inhibitor combines with the same enzyme form as the variable substrate – e.g. I_2 with A as the variable substrate, *or* if it combines with other enzyme forms upstream from the species with which the variable substrate combines and connected to that species through reversible steps. (In this context, irreversible steps are those involving either release of product at zero concentration *or addition of substrate at infinite concentration–saturation*.) Thus, in the mechanism above, the slope against $1/[B]$ is altered not only by I_1, which binds to EA, but also by I_2, because E is upstream and reversibly linked to EA. *Either of these inhibitors pulls over the enzyme distribution in such a way as to oppose saturation with A.* Of the two, only I_2, gives an intercept effect, because saturation with B reduces the steady-state concentration of EA, but not E, to zero.

Cleland states erroneously [62] that if the combination of the variable substrate with the enzyme is represented by $E' + A \rightleftharpoons E'A$, 'the inhibitor must raise E'A with respect to E' . . . . in order to affect the slope'. A dead-end inhibitor does not in fact alter the ratio of [E'A] to [E'] at all. It does, however, *decrease* the ratio of [E'A] to the *sum* of the reversibly connected species on the other side of the reaction – e.g. if $E' + I \rightleftharpoons E'I$, then $[E'A]/([E'] + [E'I])$ in the presence of inhibitor is less than $[E'A]/[E']$ in its absence.

Slope and intercept effects may also be predicted for product inhibitors. Unlike dead-end inhibitors, these act, not by siphoning off enzyme from the main reaction sequence, but rather by pushing the sequence backwards. Also the addition of a product inhibitor inevitably removes one of the 'irreversible' steps in the mechanism. Thus a product inhibitor does not always elicit the same inhibition pattern as a dead-end inhibitor combining at the same point in the mechanism. For example, in the mechanism above, a dead-end inhibitor combining with EP affects only the intercept with either A or B as the variable substrate, but Q, which also combines with EP, as a product inhibitor also affects the slopes in both cases, giving non-competitive inhibition. Its presence re-establishes the reversibility of the $EPQ \rightarrow EP$ step, and the reversal of this step works back along the line to oppose the forward conversion of E to EA and EA to EAB.

There is not space here to explore the applications of product inhibition studies, but these have been well reviewed by Cleland [62]. Each mechanism has a characteristic pattern of product inhibition, and the technique has proved of great value in choosing between mechanisms that could not be distinguished on the basis of studies of the uninhibited reaction alone. The major problems in interpretation stem from the ambiguities arising when a product inhibitor combines at more than one point in the mechanism by forming abortive complexes. Technically, also,

Table 5.6 Inhibition patterns for 3-substrate ping-pong mechanism on p.72

Dead-end inhibitor combining with:–	*Product inhibition by:–*	*Inhibition with respect to*		
		A	*B*	*C*
E		C	NC	UC
	R	C	NC	UC
EA		UC	C	UC
EAB		UC	UC	UC
F		UC	UC	C
	P	NC	NC	C
EQR		UC	UC	UC
ER		UC	UC	UC
	Q	UC	UC	NC

C = competitive, NC = non-competitive, UC = uncompetitive.

the measurements are not always easy to make, because product inhibitors tend to accentuate the curvature of reaction traces.

As an exercise in applying the qualitative rules given above, it is worth trying to write down the dead-end and product inhibition patterns for the 3-substrate ping-pong mechanism below, used as an example by Cleland [62]. The correct patterns are given in Table 5.6

$$E \overset{A}{\rightleftharpoons} EA \overset{B}{\rightleftharpoons} EAB/FP \overset{P}{\rightleftharpoons} F \overset{C}{\rightleftharpoons} EQR/FC \overset{Q}{\rightleftharpoons} ER \overset{R}{\rightleftharpoons} E$$

5.15 Isotope exchange at equilibrium

We have already encountered isotope exchange (p. 66) as a means of detecting partial reactions. Isotope exchange also provides a means of testing sequential mechanisms. The procedure and theory have been developed by Boyer and Silverstein [63, 64]. The reactants are allowed to come to chemical equilibrium in the presence of a catalytic amount of the enzyme. (Note that *all* the reactants are present in contrast to the experiments described on p. 66.) A very small amount of radioactive substrate is then added. This must not perturb the equilibrium, and, if necessary, a balancing addition of unlabelled product may be made. Under these conditions there is no *net* flow from substrates to products or vice versa, but that does not mean that the individual component reactions are at a standstill, merely that they are exactly balanced by opposing reactions. Thus, gradually, radioactivity is transferred from the labelled substrate pool to the corresponding product pool. By sampling the mixture periodically, stopping the reaction, separating the components, and measuring their level of radioactivity the rate of this exchange may be measured. Information about the mechanism is gained by studying how the rate of exchange between one substrate-product pair varies with the concentrations of the other substrate-product pair(s). We may illustrate this with the example of mitochondrial malate dehydrogenase (reaction 3, p. 11).

Steady-state studies with this enzyme, show that the mechanism is sequential. This leaves a choice between random-order and compulsory-order mechanisms. Three possibilities are illustrated below:–

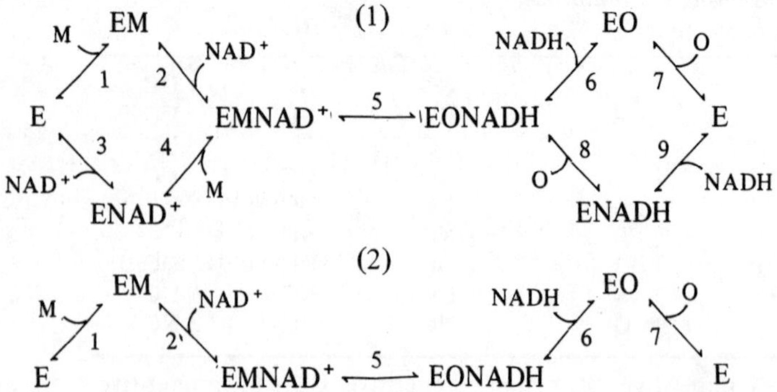

(1)

EM
M 1 2 NAD⁺
E EMNAD⁺ $\overset{5}{\rightleftharpoons}$ EONADH E
NAD⁺ 3 4 M
ENAD⁺

EO
NADH 6 7 O
O 8 9 NADH
ENADH

(2)

EM
M 1 2' NAD⁺
E EMNAD⁺ $\overset{5}{\rightleftharpoons}$ EONADH

EO
NADH 6 7 O
E

$$E \xrightarrow[NAD^+]{3} ENAD^+ \xrightarrow[M]{4} EMNAD^+ \xrightarrow{5} EONADH \xrightarrow[O]{8} ENADH \xrightarrow[NADH]{9} E \qquad (3)$$

In these three schemes the symbols M and O denote malate and oxaloacetate respectively. Let us consider the exchange of isotope from radioactive NAD^+ into NADH. In the random-order mechanism, (1), two sequences of reactions lead to exchange, the sequences. 2, 5, 6 and 3, 4, 5, 8, 9. If the concentrations of malate and oxaloacetate are low, most of the enzyme will be in the forms E, $ENAD^+$ and ENADH so that exchange occurs largely via the lower route. Its rate will be limited by Step 4, which depends on the malate concentration. If the concentrations of malate and oxaloacetate are now raised in constant ratio, so that the equilibrium is not perturbed, the rate of exchange will initially rise as Step 4 becomes faster. Progessively, however, the enzyme will be pushed over into the complexes EM, $EMNAD^+$, EONADH and EO. Thus ultimately the upper pathway takes over as the main route for exchange. Depending on the values of various rate constants; this may mean that the exchange rate rises further to a plateau, or that, after its initial increase, it declines, also to a plateau. In the second mechanism, in which coenzyme binds second, there is only one route for exchange, Steps 2, 5 and 6. At low concentrations of malate and oxaloacetate exchange would be limited by Step 1. As the concentrations of these two substrates are raised together, the exchange rate should approach a plateau value as the concentration of free enzyme diminishes. In the third mechanism, in which coenzyme leads, exchange must follow the route 3, 4, 5, 8, 9. As for the first mechanism, therefore, at very low concentrations of malate and oxaloacetate, the exchange rate should increase with the concentrations of those substrates. As the concentrations are raised still further, however, the enzyme is forced over into the central complexes, $EMNAD^+$ and EONADH. As fast as these dissociate via Steps 4 and 8, they are reformed. Thus the concentrations of $ENAD^+$ and ENADH become vanishingly small, and hence the rates of dissociation of these two complexes drop towards zero, eliminating the exchange between NAD^+ and NADH. *Raising the concentrations of the inner substrate pair ultimately eliminates exchange between the outer pair.* With pig heart mitochondrial malate dehydrogenase, [65], increasing the concentrations of malate and oxaloacetate eliminated the exchange between NAD^+ and NADH, whereas raising the concentrations of the coenzymes increased the rate of exchange between malate and oxaloacetate to a maximum plateau value. This indicates a compulsory-order mechanism with NAD^+ and NADH as the outer substrates.

6 The King and Altman procedure

Anyone who has to work out steady-state rate equations by hand, notices patterns emerging in the algebra. This is hardly surprising, since reaction mechanisms themselves contain elements of recurrent pattern and symmetry. The algebra nevertheless becomes rather daunting for complex mechanisms, through sheer volume rather than conceptual difficulty. Many short cuts are possible, but, even so, it is easy to make mistakes, dropping terms or subscripts. The problems rapidly multiply for branched mechanisms and for conditions in which the product concentrations are not all zero. For this reason the procedure of King and Altman [66] is very useful. It is based on matrix algebra and essentially simplifies the derivations by exploiting the inherent pattern to the full, and by using a shorthand notation.

Some modern books on enzyme kinetics rely heavily on the King and Altman procedure even for the treatment of simple mechanisms. I have assumed, however, that most readers of this book are likely to be unfamiliar with matrices, and are probably more concerned with understanding and using rate equations that have already been derived than with deriving new equations for novel kinetic situations. I have therefore avoided the 'magic formula' approach, and tried instead to aim for a thorough qualitative understanding of the algebra of the steady state. Against this background the King and Altman procedure may more readily be appreciated.

Let us now consider a simple mechanism:

$$E \rightleftharpoons ES \rightleftharpoons EP \rightleftharpoons E$$

and work out its rate equation by the normal algebraic route, but using in parallel the characteristic King and Altman 'arrows' so that at the end of the operation we may discover the rules of their procedure. First we must write out the mechanism in a cyclic form so that all enzyme species, including E, occur only once:

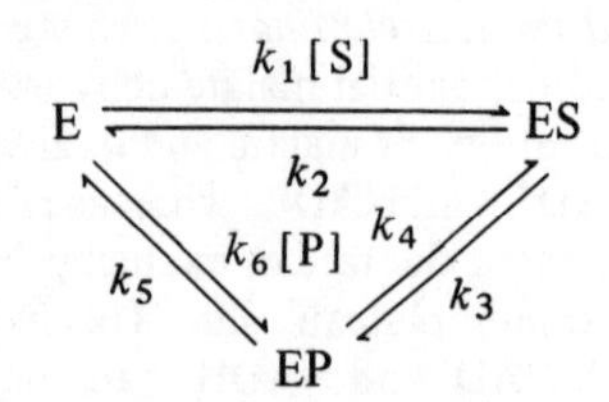

Each reaction is shown as usual by an arrow with an associated rate constant. We shall use these arrows below to symbolize the rate constants, multiplied where appropriate by associated substrate or product concentrations. Thus $\leftarrow$ simply denotes k_2, but $\rightarrow$ denotes k_1 [S]. These amended rate constants are what King and Altman term 'kappas'. This is merely a shorthand device: by replacing second-order rate constants like k_1 and k_6 by 'pseudo-first-order' rate constants k_1 [S] and k_6 [P], we ensure that the rate for each step is given by the product of two quantities only, the concentration of the relevant enzyme complex and the appropriate 'kappa'.

We can set up steady-state equations in the usual way:

For EP

$$[EP]\,(k_4 + k_5) = [ES]\,k_3 + [E]\,k_6\,[P] \tag{6.1a}$$

or, using arrows to replace the 'kappas'

$$[EP]\,(\nearrow + \nwarrow) = [ES]\,(\swarrow) + [E]\,(\searrow) \tag{6.1b}$$

For ES

$$[ES]\,(k_2 + k_3) = [EP]\,k_4 + [E]\,k_1\,[S] \tag{6.2a}$$

or

$$[ES]\,(\leftarrow + \swarrow) = [EP]\,(\nearrow) + [E]\,(\rightarrow) \tag{6.2b}$$

$\therefore$, substituting for [EP] from equation (6.1a, b),

$$[ES]\,(k_2 + k_3) = ([ES]\,k_3 + [E]\,k_6\,[P])\left(\frac{k_4}{k_4 + k_5}\right) + [E]\,k_1\,[S]$$

or,

$$[ES]\,(\leftarrow + \swarrow) = ([ES]\,(\swarrow) + [E]\,(\searrow))\left(\frac{(\nearrow)}{(\nearrow + \nwarrow)}\right) + [E]\,(\rightarrow)$$

$\therefore$, multiplying through by $(k_4 + k_5)$,

$$[ES]\,(k_2 + k_3)(k_4 + k_5) = \\ [ES]\,k_3 k_4 + [E]\,k_4 k_6\,[P] + [E]\,k_1\,[S]\,(k_4 + k_5)$$

or,

$$[ES]\,(\leftarrow + \swarrow)(\nearrow + \nwarrow) = [ES]\,(\swarrow\nearrow) + [E]\,(\searrow\nearrow) + [E]\,(\rightarrow)(\nearrow + \nwarrow)$$

(note that $[E]\,k_4 k_6\,[P]$ has been written $[E]\,(\searrow\nearrow)$ rather than $[E]\,(\nearrow\searrow)$ in order to retain the spatial relationship of the arrows in the mechanism.) We now multiply out the brackets and take across $[ES]\,k_3 k_4$ from the right hand side:–

$$[ES]\,(k_2 k_4 + k_2 k_5 + k_3 k_4 + k_3 k_5 - k_3 k_4) = \\ [E]\,(k_4 k_6\,[P] + k_1 k_4\,[S] + k_1 k_5\,[S])$$

or,

$$[ES]\,(\leftarrow\nearrow + \leftarrow\nwarrow + \swarrow\nearrow + \swarrow\nwarrow - \swarrow\nearrow) = [E]\,(\searrow\nearrow + \rightarrow\nearrow + \rightarrow\nwarrow)$$

(Note the cancellation of the $k_3 k_4$ terms, a consequence of the fact that the expressions for [EP] and [ES] both involve the k_3 and k_4 steps.)

$$\therefore \quad \frac{[\mathrm{ES}]}{k_4 k_6 [\mathrm{P}] + k_1 k_4 [\mathrm{S}] + k_1 k_5 [\mathrm{S}]} = \frac{[\mathrm{E}]}{k_2 k_4 + k_2 k_5 + k_3 k_5} \tag{6.3a}$$

or,

$$\frac{[\mathrm{ES}]}{\searrow\nearrow + \overset{\rightarrow}{\nearrow} + \overset{\rightarrow}{\nwarrow}} = \frac{[\mathrm{E}]}{\overset{\leftarrow}{\nearrow} + \overset{\leftarrow}{\nwarrow} + \nwarrow\swarrow}. \tag{6.3b}$$

Before bringing in the enzyme conservation equation, we need an expression for [EP] in terms of [E] or [ES]. This can be readily obtained by substituting into equation (6.1) the expression (equation (6.3)) for [E] in terms of [ES], but there is an even easier way. The overall mechanism is symmetrical if we remember to use 'kappas' rather than unamended rate constants. Thus, if we have an equation relating [ES] and [E], we should be able to use this symmetry to write down *by inspection* a similar equation relating [EP] and [ES]. This is where the arrows come in useful, because they now help us to find the symmetrical relationship very quickly. If you examine the denominator of the left hand side of equation (6.3b), you will see that the three pairs of arrows *all end at ES* in the mechanism, and that these are the *only such pairs*, and that there are no opposing arrows (e.g. $\nearrow\swarrow$). Similarly on the other side of the equation are the three pairs of arrows that end at E. It should be obvious that we may now add a third equality to equation (6.3):

$$\frac{[\mathrm{ES}]}{\searrow\nearrow + \overset{\rightarrow}{\nearrow} + \overset{\rightarrow}{\nwarrow}} = \frac{[\mathrm{E}]}{\overset{\leftarrow}{\nearrow} + \overset{\leftarrow}{\nwarrow} + \nwarrow\swarrow} = \frac{[\mathrm{EP}]}{\overset{\leftarrow}{\searrow} + \searrow\swarrow + \overset{\rightarrow}{\swarrow}}$$

or

$$\frac{[\mathrm{ES}]}{k_4 k_6 [\mathrm{P}] + k_1 k_4 [\mathrm{S}] + k_1 k_5 [\mathrm{S}]} = \frac{[\mathrm{E}]}{k_2 k_4 + k_2 k_5 + k_3 k_5}$$

$$= \frac{[\mathrm{EP}]}{k_2 k_6 [\mathrm{P}] + k_3 k_6 [\mathrm{P}] + k_1 k_3 [\mathrm{S}]}.$$

If you are unconvinced of the validity of this argument based on symmetry, work through the algebra as above, but using the steady-state equations for E and ES rather than EP and ES. Abbreviating the denominators in equation (6.3) as D_1, D_2 and D_3, we may write:–

$$\frac{[\mathrm{ES}]}{D_1} = \frac{[\mathrm{E}]}{D_2} = \frac{[\mathrm{EP}]}{D_3}. \tag{6.4}$$

We now introduce the enzyme conservation equation:–

$$e = [\mathrm{E}] + [\mathrm{ES}] + [\mathrm{EP}]$$

$\therefore$ substituting for [E] and [ES] from equation (6.4)

$$e = [\mathrm{EP}]\left[\frac{D_2}{D_3} + \frac{D_1}{D_3} + 1\right] = \frac{[\mathrm{EP}](D_1 + D_2 + D_3)}{D_3}$$

$$\therefore \qquad [EP] = \frac{eD_3}{D_1 + D_2 + D_3} \tag{6.5}$$

From equation (6.4)

$$[E] = \frac{eD_2}{D_1 + D_2 + D_3}. \tag{6.6}$$

Thus for each complex in the mechanism there is an expression (D_1 for ES etc.) which, when divided by the sum of all such expressions, gives the fraction of the enzyme present as that complex in the steady state. As we have seen, the expression for a given complex may readily be written down by inspection as *a sum of arrow patterns all ending at that complex*. For more complicated mechanisms the expressions become larger, and each term includes more rate constants: for an unbranched mechanism with *four complexes*, the corresponding expressions would contain *four terms*, each the product of *three rate constants*. For such a mechanism it is much quicker to write down the distribution equations in this way than to work through all the algebra. This is the King and Altman procedure in essence.

Once the distribution equations are written out as in equations (6.5) and (6.6), the net rate in the forward direction is given by:—

$$v = [EP]k_5 - [E]k_6[P]$$

$$(= [ES]k_3 - [EP]k_5 = [E]k_1[S] - [ES]k_2)$$

$$= \frac{e(k_5 D_3 - k_6 [P] D_2)}{D_1 + D_2 + D_3}. \tag{6.7}$$

The next and final stage is to expand and regroup the terms:—

$$\frac{v_0}{e} = \frac{k_2k_5k_6[P] + k_3k_5k_6[P] + k_1k_3k_5[S] - k_2k_4k_6[P] - k_2k_5k_6[P] - k_3k_5k_6[P]}{k_2k_4 + k_2k_5 + k_3k_5 + k_1(k_3 + k_4 + k_5)[S] + k_6(k_2 + k_3 + k_4)[P]}$$

$$= \frac{k_1k_3k_5[S] - k_2k_4k_6[P]}{k_2k_4 + k_2k_5 + k_3k_5 + k_1(k_3 + k_4 + k_5)[S] + k_6(k_2 + k_3 + k_4)[P]}. \tag{6.8}$$

This equation can be simplified to give the initial rate in the absence of product by setting [P] = 0, or the initial rate in the reverse direction in the absence of substrate by setting [S] = 0.

Recently, attempts have been made to simplify this procedure still further, relating it more obviously and immediately to the structure of the mechanism [e.g. 67, 68]. On the basis of these simplified procedures, programmes have been written [69, 70] which make possible computer derivation of initial-rate equations [60].

7 Non-linear kinetics and the concept of allosteric interaction

7.1 Introduction

We have so far considered only 'linear' systems, i.e. enzymes that give hyperbolic plots of v against [S] and linear plots of $1/v$ against $1/[S]$. Although many enzymes do fall into this category, others clearly deviate from linearity within the normally studied ranges of substrate concentration, and these include many enzymes occupying strategically important positions in metabolism. The non-linearity can frequently be rationalized teleologically in terms of desirable control characteristics.

Although metabolic regulation and the control of enzyme activity are covered elsewhere in the series [71,72] this book would be incomplete without some discussion of non-linear kinetics. The available models can be most readily assessed in the context of the kinetic theory of simpler systems.

7.2 Allostery and its antecedents

In 1963, Monod, Changeux and Jacob published a paper [73] entitled 'Allosteric Proteins and Cellular Control Systems'. In it they proposed that in addition to active, substrate-specific sites, regulatory enzymes might possess separate 'allosteric' sites specific for their regulators. Binding of a regulator molecule at such a site, they suggested, could influence events at the active site. The idea was not entirely new, but the paper drew together the information then available to provide a clear, persuasive argument for the widespread occurrence of regulation by this means. The name 'allosteric' was intended to emphasize that the regulator need not bear any structural resemblance to the substrate.

Let us examine some of the strands of evidence that led to this formal statement of the concept of allosteric control. The first strand, historically, came from work on haemoglobin – not an enzyme but a 'binding protein' for oxygen. The difference in oxygen concentration between the lungs and other tissues is not very great, and yet to function efficiently haemoglobin needs to be almost saturated with oxygen in the lungs and nearly completely discharged in the tissues. To achieve this, it has a sigmoid saturation curve (Fig. 7.1) instead of the usual hyperbola. At low oxygen concentrations the fractional saturation of the carrier rises very slowly, and then, over a relatively small range, the physiological range, the oxygen sites nearly all become filled. A useful measure of this sigmoidicity is R_S, the 'cooperativity index', defined as the ratio between the substrate or ligand concentrations required for 90% and 10% saturation [74]. For simple hyperbolic binding $R_S = 81$ (from equation

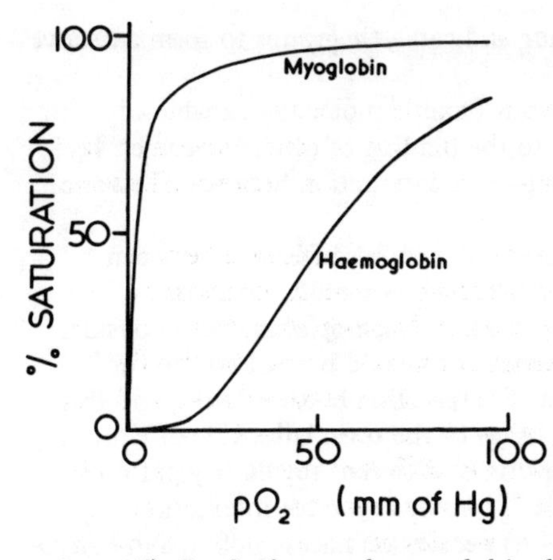

Fig. 7.1 A comparison of the oxygen dissociation curves of the closely related proteins myoglobin and haemoglobin.

2.5, p. 17). R_S for human haemoglobin, however, is about 10.

The successive analyses of Hill [75], Adair [76] and Pauling [77] led to the view that four oxygen sites per haemoglobin molecule interact in a cooperative fashion, so that filling of one facilitates filling of the next. The X-ray structure shows however, that the four haem groups are widely separated [78]. Any interaction must therefore be a long-range interaction.

The second strand came from metabolic biochemistry. Studies of bacterial biosynthesis [79, 80, 81, 82] and of mammalian glycogen phosphorylase [83] revealed that the end-product of a metabolic pathway often inhibits the first specific step of a metabolic pathway, thereby minimizing waste. The 'feedback' inhibitors are often very different in structure from the substrates of the regulated enzymes, making it difficult to visualise the process in terms of competition for a common site. In many cases the reaction rate shows a sigmoid dependence on the substrate concentration. Studies of the regulation of aspartate transcarbamylase by CTP [84], and of threonine deaminase by isoleucine [85] led in both cases to the proposal of separate feedback sites. Again, therefore, the suggestion of long-range interactions.

The abstract idea of interacting sites demanded a physical mechanism. In the early days of enzymology, the remarkable substrate specificity of enzymes had led to the picture of a rather rigid structure, the 'lock' which could only be fitted by the corresponding 'key', the substrate. Physical studies showed, however, that proteins might be more flexible. Many enzymes could be reversibly unfolded, and smaller, local conformational changes could be induced by small molecules not necessarily in any way related to the substrate. With the 'Induced Fit' theory [86], Koshland went a step further, suggesting that an enzyme's function may sometimes *depend* on its flexibility. For some hydrolases acting on large substrates, smaller analogues which apparently should 'fit the lock', fail to act as substrates. Thus in some cases the key must complete its own lock, displacing any blocking groups and gathering around itself the necessary binding and catalytic groups to form an active enzyme-substrate complex.

Here, in the picture of proteins as flexible molecules capable of subtle conformational responses to the binding of other molecules, lay also a possible mechanism for long-range interaction between allosteric sites.

The original emphasis [73] on the structural difference between substrate and regulator is rather misleading; it would, for instance, exclude the archetypal regulatory protein, haemoglobin, from consideration as a case of allostery! The emphasis should be, not on the site specificity, but rather on the fact of interaction between sites, and this has indeed become the accepted usage of the term 'allosteric effect'.

To allow unambiguous description of different regulatory patterns, two adjectives were introduced [87] : *homotropic* for interactions between identical ligands, e.g. O_2 molecules on haemoglobin, and *heterotropic* for effects of one compound upon the binding of another, e.g. the effect of CTP on aspartate binding by aspartate transcarbamylase. (The latter term does not include such non-allosteric effects as competition for a single site.)

7.3 Does non-linear kinetics necessarily imply allosteric interaction?

Monod, Changeux and Jacob warned that, 'The most serious objection to the concept of allosteric control is that it *could* be used to "explain away" almost any mysterious physiological phenomenon'. So it proved to be. Armed with the new talisman, the biochemists of the '60's postulated allosteric sites in profusion for most regulatory enzymes. The vogue inevitably provoked a reaction. Most of the candidates for allostery were in fact enzymes with two or more substrates. It was already known [88, 89] that the steady-state equation for a random-order two-substrate mechanism contains squared terms in each substrate concentration, predicting non-linear kinetics (see Chapter 5, p. 60). Ferdinand [90] showed that, if:–

$$v = \frac{i[S] + j[S]^2}{k + l[S] + m[S]^2}$$

then the plot of v against [S] is sigmoid if $k/l > i/j$. The possibilities for sigmoid kinetics inherent in steady-state two-substrate mechanisms were also explored by Sweeny and Fisher [91].

Another model, due to Rabin, (Fig. 7.2) requires neither an allosteric site nor a second substrate [92]. It postulates instead two conformational states of the enzyme, E and F, E being preferred in the absence of substrate. Both forms bind substrate, but only F catalyses the reaction. Product release yields free enzyme in the F form. At high [S], this is immediately re-converted to FS. At lower [S] values, however, F has a chance of isomerising to E. A significant portion of the flow then proceeds via ES, the rate of product formation being limited by the

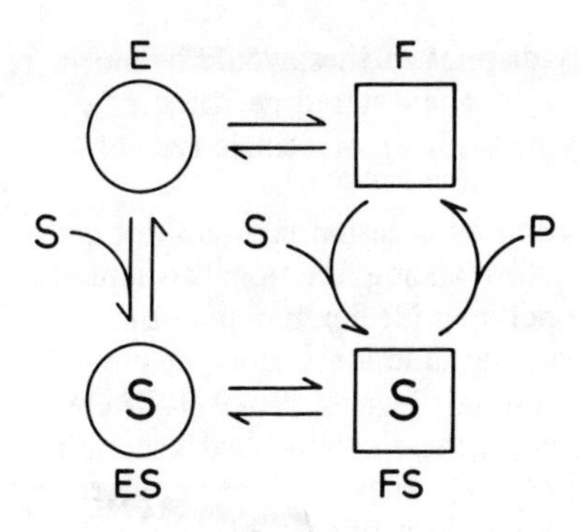

Fig. 7.2 Rabin's model to explain substrate cooperativity without postulating allosteric sites.

isomerization of ES to FS. Steady-state analysis of this model also gives a rate equation with terms in $[S]^2$ which could explain sigmoid kinetics.

A third possible non-allosteric explanation for kinetic or binding curves that depart from the simple classical pattern is the presence of non-identical sites for the substrate or ligand [93]. A mixture of two isoenzymes, for instance, will not give 'linear' kinetics unless the two forms have the same K_m. This may be verified by summing two Michaelis–Menten expressions (see equation 2.5) with different values of K_m. Non-identity may also arise in more subtle ways: proteins may have identical subunits arranged in non-equivalent positions. There might be, for instance, distinguishable 'top' and 'bottom' sets of subunits; the behaviour of one set might be altered by the mere presence of the other set, if, for example, one set of binding sites were near the interface between the two sets of sub-units.

The sites with greatest affinity would, however, bind the ligand preferentially, and thus non-identical sites can only be invoked in cases of apparent 'negative cooperativity' – i.e. where initial binding appears to impede subsequent binding.

7.4 Does allosteric interaction require a multi-subunit protein?
Non-linear kinetics, as we have seen, is not in itself conclusive evidence of allosteric interaction. Nevertheless, for many enzymes there is abundant additional evidence of communication between separate sites. The general statement of Monod, Changeux and Jacob was followed by attempts to formulate more explicit models. The evidence then available suggested that allosteric interaction occurred only in multi-subunit (oligomeric) proteins. There was the obvious contrast between tetrameric haemoglobin and related myoglobin, monomeric and not allosteric (Fig. 7.1). Again, the separate regulatory and catalytic sub-units found in aspartate transcarbamylase [94] clearly implied subunit interaction. There seemed to be also a compelling evolutionary argument for oligomeric structure as a prerequisite for allosteric behaviour. To quote Monod, Wyman and Changeux [87], commenting on the case of aspartate transcarbamylase, 'The emergence and evolution of such

structures, by association of primitively distinct entities, would be much easier to understand than the acquisition of a new stereospecific site by an already existing and functional enzyme made up of a single type of subunit.'

Since then we have learnt that evolution on occasion takes a short cut by 'stitching' together two genes coding for existing functional structures to make a new gene coding for a single polypeptide combining both functions. Many proteins have more than one functional 'site' per subunit, some even two different *active* sites per subunit [e.g. 95, 96, 97]. There are also enzymes such as glutamate dehydrogenase in which each subunit bears both active and regulatory sites [e.g. 98]. There have been reports, moreover, of allosteric behaviour in monomeric enzymes [99, 100]. Conformational flexibility is certainly not restricted to oligomeric proteins. It is no longer obvious, therefore, that allosteric behaviour implies a multi-subunit structure. It is important to bear this in mind in considering the models below.

7.5 The main allosteric models

Two models have dominated the discussion of allosteric enzymes in recent years. They are due to Monod, Wyman and Changeux [87] and to Koshland, Némethy and Filmer [74]. The relationship between them can be appreciated by reference to the more general 'parent model' [101], illustrated in Fig. 7.3. They have several features in common:–

(i) in both models the allosteric effects are mediated through interactions between neighbouring subunits in an oligomer. For simplicity we shall consider the case of a dimer (Fig. 7.3);

(ii) in both models two distinct conformational states are available to each subunit. These are shown in Fig. 7.3 as ○ and □. Koshland *et al.* refer

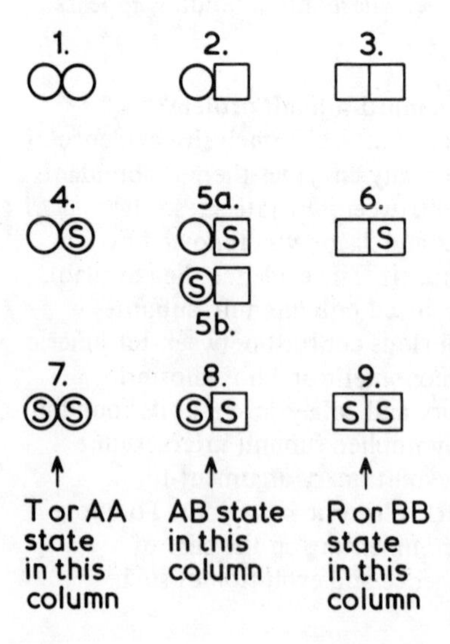

Fig. 7.3 A general model for substrate binding and conformational change in a protein with two identical subunits and two conformational states. Conformational change proceeds from left to right from the all-circle (T or AA) state on the left to the all-square (R or BB) state on the right. Substrate saturation proceeds from the top row to the bottom row. The numbers have been added merely for ease of reference in the text.

to these as the A and B states, while Monod *et al.* call them the T and R states:

(iii) both models are concerned first and foremost with binding. It is assumed that the rate of any reaction subsequent to binding is directly proportional to the extent of saturation. Monod *et al.* [87] do mention 'V systems', in which the rate of product formation from the ES complex is affected by the allosteric transition, but they dwell mainly on 'K systems', in which only the affinity of E for S is altered. This preoccupation with ligand binding is a consequence of the historical position of haemoglobin.

In Fig. 7.3, increasing [S] inevitably displaces the enzyme distribution towards the bottom row(s). If both conformational forms, ○ and □, have identical affinity for S, then a simple hyperbolic saturation curve is to be expected. Likewise, if in the absence of S the protein is almost entirely in one conformational state, say □, and S binds as tightly to this form of the protein as to the ○ state, the latter will never be present to a significant extent, and the saturation curve will again be hyperbolic.

By contrast, in the *Monod, Wyman and Changeux (MWC) model* it is assumed (a) that the conformational equilibrium in the absence of the ligand lies over in favour of the form that binds the ligand *least* well, the T form; (b) that hybrid states (the middle column in Fig. 7.3) are virtually absent. This implies that the interaction between a square and a circle must be energetically very much less favourable than between two circles or two squares. Because the protein is assumed to be stable only in all-square or all-circle states, the MWC model is often referred to as the "symmetry model".

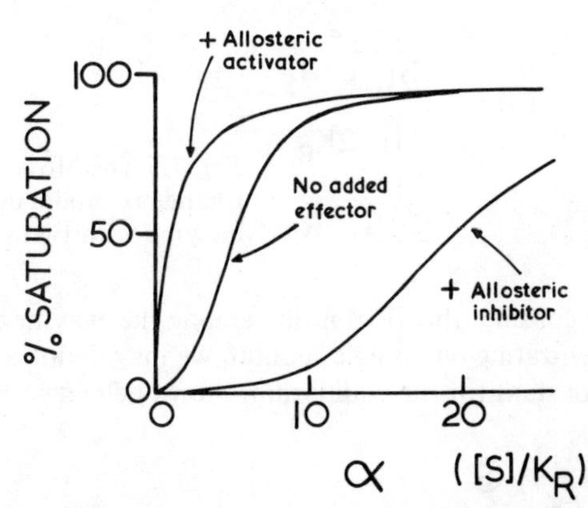

Fig. 7.4 Positive homotropic cooperativity in substrate binding. The middle curve shows the cooperative binding of a substrate, plotted as a function of α, the ratio of [S] to K_R, the constant governing dissociation of S from the R form of the enzyme in the MWC model. The upper curve shows that allosteric activators tend to diminish the homotropic cooperativity, whilst the lower curve shows that allosteric inhibitors tend to enhance the homotropic cooperativity.

When the ligand, S, is added it tends to pull the equilibrium over towards the form (the R form, □) that binds it most strongly. Since a square conformation in a filled subunit enforces the tight-binding conformation on the adjacent vacant subunit(s), the second site is more easily filled than the first. Thus S facilitates its own binding. This is *positive homotropic cooperativity* and results in a sigmoid saturation curve (Fig. 7.4). The degree of sigmoidicity in the MWC model depends on the relative affinities of the R and T forms for S, and on the extent to which the T state is favoured in the absence of S.

Since S cannot displace the conformational equilibrium in favour of a form which binds it less tightly, *the MWC model cannot produce negative homotropic cooperativity*. This is an important distinction between the MWC model and that of Koshland *et al.* (KNF).

To derive a rate equation for an enzyme obeying the MWC model, we must find out how the fractional saturation of the enzyme varies with [S]. We shall refer to Fig. 7.5, which incorporates the assumption of compulsory symmetry. Note that the rapid equilibrium between the T and R states is described by a single *allosteric constant*, *L*. The value of *L* must depend both on the intrinsic stability of a ○ relative to a □, and on the strength of interaction between two ○ subunits relative to that between two □'s. Unlike the KNF model, however, the MWC model does not distinguish the separate contributions of these two factors.

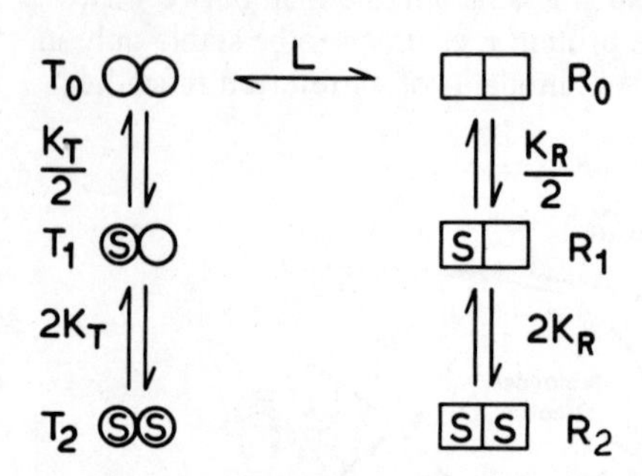

Fig. 7.5 The Monod–Wyman–Changeux model depicted for an enzyme with two subunits.

Let us first consider the distribution among the enzyme forms T_0, T_1 and T_2. Concentrating on a single subunit, we may define a 'microscopic' dissociation constant for the equilibrium shown below:–

$$\bigcirc \underset{k_2}{\overset{k_1}{\rightleftharpoons}} Ⓢ$$

$$K_T = \frac{[\bigcirc][S]}{[Ⓢ]} = \frac{k_2}{k_1}$$

For the dissociation of T_1 to T_0, and S, however, the apparent dissociation constant is not simply K_T. T_1 can only release one molecule of S, but T_0 has *two* subunits, each able to bind S. The dissociation constant for this step is therefore $K_T/2$. Similarly the dissociation constant for interconversion of T_1 and T_2 is $2K_T$.

We can now write out equations for the distribution of the enzyme among various forms:–

$$[R_1] = [R_0]\frac{2[S]}{K_R} = [R_0]2\alpha, \quad \text{where } \alpha = \frac{[S]}{K_R}$$

$$[R_2] = [R_1]\frac{[S]}{2K_R} = [R_0]\left(\frac{[S]}{K_R}\right)^2 = [R_0]\alpha^2$$

$$[T_0] = L[R_0]$$

$$[T_1] = [T_0]\frac{2[S]}{K_T} = L[R_0]\frac{2[S]}{K_T} = L[R_0]2\alpha c$$

where $c = K_R/K_T$.

$$[T_2] = [T_1]\frac{[S]}{2K_T} = [T_0]\left(\frac{[S]}{K_T}\right)^2 = L[R_0]\alpha^2 c^2$$

The total number of subunits is $2([R_0] + [R_1] + [R_2] + [T_0] + [T_1] + [T_2])$. The number of *filled* subunits is $[R_1] + 2[R_2] + [T_1] + 2[T_2]$. The fractional saturation, $\bar{Y}$, is therefore given by equation (7.1):–

$$\bar{Y} = \frac{[R_1] + 2[R_2] + [T_1] + 2[T_2]}{2([R_0] + [R_1] + [R_2] + [T_0] + [T_1] + [T_2])} \tag{7.1}$$

∴ Substituting from above,

$$\bar{Y} = \frac{[R_0](2\alpha + 2\alpha^2 + 2L\alpha c + 2L\alpha^2 c^2)}{2[R_0](1 + 2\alpha + \alpha^2 + L + 2L\alpha c + L\alpha^2 c^2)}$$

$$= \frac{\alpha(1+\alpha) + L\alpha c(1 + \alpha c)}{(1+\alpha)^2 + L(1+\alpha c)^2} \tag{7.2}$$

In the original paper [87] the equation is worked out for the general case of a protein with n sites:–

$$\bar{Y} = \frac{\alpha(1+\alpha)^{n-1} + L\alpha c(1+\alpha c)^{n-1}}{(1+\alpha)^n + L(1+\alpha c)^r} \tag{7.3}$$

When $c = 1$ or $L = 0$, this equation reduces to the Michaelis–Menten form. When, on the other hand, $c = 0$ i.e. when S can only bind to the R form, it simplifies to:–

$$\bar{Y} = \frac{\alpha(1+\alpha)^{n-1}}{L + (1+\alpha)^n}. \tag{7.4}$$

Thus sigmoidicity is more marked when L is large and/or when c is small.

Monod *et al.* [87] show saturation curves for various combinations of these constants in the case of a tetrameric protein.

Given this basis of homotropic cooperativity, the action of heterotropic effectors is explained by displacement of the conformational equilibrium. A compound binding more strongly to the R state at an allosteric site would be an activator and would make the saturation curve for S *less* sigmoid (Fig. 7.4). Conversely, an allosteric inhibitor would act by stabilising the T state, making the saturation curve for S *more* sigmoid.

In the *KNF model* there is no assumption of symmetry; hybrid states are allowed. The characteristics of the saturation curve, according to this model, are determined by the relative strengths of the subunit interactions in various states of the oligomer, and by the differing affinities of the A and B subunits for the ligand.

Koshland *et al.* approached allosteric interaction from the standpoint of 'Induced Fit' (p. 79), in which the emphasis is on the ligand as the primary causative agent of conformational change. This has led to some confusion. In fact, Koshland *et al.* explicitly recognized [74] that a ligand might bring about conformational change by displacing a pre-existing equilibrium rather than by enforcing the change *de novo.* They were primarily concerned, however, with the *distribution* among possible enzyme species at equilibrium; the phenomena that the KNF and MWC models seek to explain are independent of the route(s) taken to reach that distribution. The confusion has arisen because, in the simplest and most widely-discussed version of the KNF model, it is assumed a) that the B conformation (□) is found only in subunits that contain S– i.e., that such species as 2,3,5b and 6 in Fig. 7.3 never form a significant proportion of the equilibrium mixture; b) that only □ subunits can bind the ligand; this 'exclusive binding' deletes species 4, 7 and 8 and corresponds to the assumption for the MWC model that $c = 0$ (equation 7.4). This leaves only the species lying on the diagonal in Fig. 7.3., i.e. 1, 5a and 9. This unfortunately has become known as the 'simple sequential model', a name that clearly implies the *route* of conformational change. The low equilibrium concentration of the off-diagonal species does *not*, however, rule them out as intermediates in the interconversion of 1, 5a and 9.

A mistaken emphasis on the route of conformational change, together with the use of different terminology, has tended to exaggerate the difference between the MWC and KNF models. The real distinction is quantitative rather than qualitative, lying in different relative values of equilibrium constants rather than in absolute prohibition of various routes. In what follows, therefore, instead of confining ourselves to any single restrictive form of the KNF model, we shall use the KNF approach and terminology to discuss the general 'parent mechanism' [101] (Fig. 7.3). In this mechanism none of the species are deleted, all being present in proportions that depend on [S] and on the values of the various constants.

To allow scope for allosteric interaction we assume as before that B subunits (□) bind the ligand more tightly than A subunits (○). Let us also assume that *in the monomer* (whether or not it is actually present) the A state is intrinsically more stable than the B state. Expressing this in KNF terminology the transformation constant $K_t < 1$, where $K_t = [B]/[A]$. The actual distribution among the three possible *dimeric* states (Fig. 7.3) in the absence of S depends in addition on the subunit interactions – e.g. if an A (○) interacts more favourably with a B (□) than with another A, this counteracts the intrinsic instability of B subunits, and AB dimers (○□) may form a significant fraction of the total protein.

Note the separation of the *intrinsic* stability of the two monomeric states from the contribution of the interactions between them. The latter factor is quantitatively described by the use of three stability constants, K_{AA}, K_{AB} and K_{BB}.

$$K_{AA} = \frac{[AA]}{[A]^2}; \qquad K_{AB} = \frac{[AB]}{[A]\,[B]}; \qquad K_{BB} = \frac{[BB]}{[B]^2}$$

$$\therefore \quad \frac{K_{AB}}{K_{AA}} = \frac{[AB]\,[A]}{[AA]\,[B]} = \frac{[AB]}{[AA]}\,\frac{1}{K_t} \tag{7.5}$$

$$\text{and} \quad \frac{K_{BB}}{K_{AA}} = \frac{[BB]\,[A]^2}{[AA]\,[B]^2} = \frac{[BB]}{[AA]}\,\frac{1}{K_t^2}. \tag{7.6}$$

For a discussion of the potentialities of the model what matters is the *relative* values of these constants. Accordingly Koshland *et al.* assigned to K_{AA} an arbitrary value of 1. It follows from equations (7.5 and 7.6) that

$$K_{AB} = \frac{[AB]}{[AA]}\,\frac{1}{K_t} \qquad \text{and} \qquad K_{BB} = \frac{[BB]}{[AA]}\,\frac{1}{K_t^2}.$$

The division here by K_t and K_t^2 corrects for the intrinsic stability of the two monomeric states, leaving only the contribution of the subunit interactions.

If the interactions between A and B are similar to those between A and A, but much less favourable than between B and B ($K_{AA} = K_{AB} \ll K_{BB}$), positive cooperativity results. Filling the first site tends to force the filled subunit into the B conformation, because B binds S more tightly than A. This in turn converts vacant subunits to the B conformation. Thus many more vacant, tight-binding B subunits become available as [S] is raised, so that the apparent dissociation constant for the second site is lower than for the first. Referring to Fig. 7.4, the cooperativity can be attributed to the fact that, although very little of Species 2 is formed, significant amounts of Species 6 are formed as [S] is raised.

Conversely, if $K_{AA} = K_{AB} \gg K_{BB}$, there is *negative* cooperativity. Whatever the concentration of S required to fill the first site, a much higher concentration is needed to overcome the protein's reluctance to assume the all-square BB state. Thus the second site on the dimer is more difficult to saturate than the first.

If, on the other hand, $K_{AA} > K_{BB} \gg K_{AB}$, we get the MWC model as a special case of the general parent model, because this condition ensures that the hybrid dimer cannot form a significant proportion of the equilibrium mixture. Comparison with the previous condition, for negative cooperativity, makes it clear why the MWC model cannot give rise to such behaviour.

The equilibrium distribution between the unliganded species 1, 2 and 3 in Fig. 7.3 is given by equations (7.5 and 7.6). To obtain the full saturation function we need also the dissociation constants for the binding of S to monomers A and B. The distribution down each column in Fig. 7.3 can be worked out as for the MWC model (p. 84) by using these constants together with the appropriate statistical factors. The enzyme conservation equation is then applied as usual to give the saturation function.

So far, we have only considered interactions in a dimer. The literature concentrates on tetramers; haemoglobin is a tetramer, and both Monod *et al.* and Koshland *et al.* were concerned to show that their models could fit the oxygen binding curve of haemoglobin convincingly. If hybrid conformational states can form a significant fraction of the total protein, then in a multi-subunit protein, the actual arrangement of the subunits becomes important. Thus Koshland *et al.* [74] devote a good deal of attention to the permutations of subunit interactions in various geometric arrangements of a tetramer – square, linear, tetrahedral. Such detailed considerations are not, however, essential to a general understanding of the Koshland approach. In the MWC model the issue does not arise because of the symmetry assumption.

7.6 Are the allosteric models adequate?

Even the most general model discussed above is still clearly restrictive in several respects. It postulates only two conformational states. One should not postulate more without reason, but the possibility should be borne in mind. The model assumes identical subunits. This is the most common situation, but there are exceptions. When applied to enzymes, the model uses the equilibrium assumption, which is arbitrary and only occasionally valid (Chapter 2). Scant consideration is given to the allosteric alteration of catalytic constants.

The MWC and KNF models, therefore, are starting points. They provide a conceptual framework and a language with which to tackle the problems of allosteric proteins. In individual cases one must be ready to invoke whatever extra variables may be required. This is not a criticism of the basic models. It would be unreasonable to expect a single model to embrace all the subtly different means of control employed in different proteins.

7.7 Experimental tests of allosteric models.

The first approach to the testing of allosteric models comes under the general heading of *'curve fitting'.* Any valid model must be capable, with suitably chosen constants, of generating an equation that closely fits the experimental points. For some enzymes it is easy to rule out the MWC model on this basis Bovine glutamate dehydrogenase [102], rabbit glyceraldehyde 3-phosphate dehydrogenase [103, 104] and CTP synthetase from *E. coli* [105], for example, give kinetic plots and/or saturation curves too complex to be explained by the MWC model, and it is necessary to postulate negative or mixed [106] binding cooperativity. In at least one of these cases it seems necessary to invoke interactions affecting the catalytic rate constants [107]. In general, however, the analysis of complex curve shapes places a heavy demand on experimental precision, and may require measurements over an impracticably wide range of substrate concentration. Simple curves can often be fitted equally well by different mechanisms; the saturation curve of haemoglobin can be fitted by either the MWC model [87] or the 'simple sequential' version of the KNF model [74].

Another approach depends on the measurement of *'relaxation times'.* In the presence of a ligand a protein establishes equilibrium among the various possible liganded and unliganded species. If the temperature is suddenly raised by a few degrees in a 'temperature jump' apparatus, the equilibrium is perturbed and has to readjust. With suitable optical methods and rapid recording facilities this readjustment may be monitored. It is usually possible to break down the re-equilibration into a series of pseudo-first-order processes, each one corresponding to one of the equilibrium reactions making up the overall system. The 'relaxation times' are the half-times of these pseudo-first-order processes. For the MWC model one would expect to see a maximum of three relaxation times, one independent of substrate concentration for the allosteric transition, and two substrate-dependent relaxation times for the binding of S to the R and T states. By contrast, for the 'simple sequential' model one would predict four relaxation times, because each binding step would have a different equilibrium constant.

Temperature jump studies of the binding of NAD^+ by yeast glyceraldehyde 3-phosphate dehydrogenase [108] have yielded results consistent with the MWC model. One must be cautious, however, in interpreting such results. Individual constants may escape detection, either by being below the time resolution of the instruments, or by being similar in magnitude to other constants in the mechanism so that there is an overlap. Furthermore, in rapid-reaction studies the route through the mechanism *cannot* be ignored (cf. p. 86). For instance, in the MWC mechanism the allosteric transition, shown as a single arrow in Fig. 7.5, may actually proceed in two or more kinetically distinct steps: intermediate species that may be insignificant contributors to the total at equilibrium may nevertheless become temporarily significant during the re-establishment of equilibrium. In the specific case of glyceraldehyde 3-phosphate dehydrogenase from yeast, one must bear in mind also that the corresponding enzyme from rabbit muscle [103, 104] clearly does not follow the MWC model. It seems hardly likely that the allosteric mechanisms for two related forms of the same enzyme would be totally different. This reinforces the point made earlier, viz. that the simplified allosteric models should be regarded merely as extremes of the same general model, and not as fundamentally distinct mechanisms.

Recent temperature jump studies of dimeric yeast hexokinase [109] have shown that neither the MWC model nor the simple sequential version of the KNF model accounts for the behaviour of this enzyme. The full, unabridged 'parent model' (p. 82) seems to be required.

Probably the most reliable way of testing allosteric mechanisms at present is to measure conformational change as a function of the degree of saturation with the ligand. This requires an independent indicator of conformation – either a sensitive property of the protein, such as fluorescence, circular dichroism, chemical reactivity etc., or a specific attached 'reporter group'. Thus Ogawa and McConnell [110] used a spin label (i.e. a chemical reporter group with an e.p.r. signal due to an unpaired electron) to study haemoglobin. They detected only two conformational states, and found that the extent of conformational change was proportional to the degree of saturation with oxygen. One would expect strict proportionality for the 'simple sequential' model, whereas for the MWC model one might expect the extent of conformational change to run ahead of the extent of saturation. Similar arguments have been used to support a concerted transition of the MWC type for E. coli aspartate transcarbamylase [111], but this conclusion has been contested [112].

Finally, mention must be made of the Hill equation [75] and the Hill plot, first introduced for analysis of the oxygen dissociation curve of haemoglobin. If the binding of ligand by a protein with several sites can be adequately described by the following equation

$$E + nS \rightleftharpoons E(S)_n$$

then the equilibrium expression is given by

$$[E(S)_n] = [E]K[S]^n \tag{7.7}$$

where K is the equilibrium constant. Rewriting, we have:–

$$\frac{[E(S)_n]}{[E]} = K[S]^n.$$

If the fractional saturation (i.e. the fraction of the total number of binding sites that is filled) is given the symbol $\bar{Y}$, then the fraction of unfilled sites is $1 - \bar{Y}$. It follows that

$$\frac{\bar{Y}}{1-\bar{Y}} = K[S]^n.$$

Taking logarithms, we obtain the Hill equation,

$$\log \frac{\bar{Y}}{1 - \bar{Y}} = \log K + n \log[\mathrm{S}]. \tag{7.8}$$

If the initial assumption is correct, therefore, a plot of log $(\bar{Y}/(1 - \bar{Y}))$ against log[S] should give a straight line with a slope of n and an ordinate intercept of log K. The underlying postulate is one of extremely strong positive cooperativity i.e., for a dimer, that only species 1 and 9 in Fig. 7.3 are significant.

Barcroft and Hill found the value of n for haemoglobin to be 2.8 from the Hill plot. We now know that the true number of binding sites is 4, and can conclude that the Hill model is inadequate for haemoglobin. Nevertheless, the plot has continued to be used for other proteins. Many enzymes do give roughly linear Hill plots over a limited range of substrate concentration, and since an enzyme obeying simple Michaelis–Menten kinetics should give a slope of 1 in the Hill plot, higher slopes are taken to be an approximate measure of the degree of cooperativity. In at least one case the Hill coefficient (the slope of the plot) has been found to be equal to the number of binding sites. This is for deoxy CMP deaminase from donkey spleen: in the presence of its allosteric inhibitor, deoxy TTP, the sigmoidicity of the rate dependence on deoxy CMP is increased, and the Hill plot gives a slope of 4 [113].

7.8 Prospects

We are on the brink of rapid advances in our understanding of allosteric mechanisms. X-ray structures are now revealing, not only subunit arrangements, but also the nature of conformational changes. Crystallography, however, as a static technique, cannot supplant kinetics. It can give us a very clear picture of the 'before' and the 'after' but not the 'during'! There can be no doubt that kinetic studies will continue to play a central role in answering the many tantalising questions about enzyme action that remain.

References

[1] Hartley, B.S. (1974), *Soc. Gen. Microbiol. Symp.* **24**, 151–182.

[2] Brown, P.R. and Clarke, P.H. (1972), *J. Gen. Microbiol.*, **70**, 287–298.

[3] Bergmeyer, H.–U. (1974), *Methods of Enzymatic Analysis* (2nd English edn.) Vol. 1, pp. 3–92, Verlag Chemie, Weinheim.

[4] Wiseman, A. (ed.) (1975), *Handbook of Enzyme Biotechnology*, Ellis Horwood, Chichester.

[5] Bachelard, H.S. (1974), *Brain Biochemistry*, pp. 34–36, Chapman and Hall, London.

[6] Tempest, D.W., Meers, J.R. and Brown, C.M. (1971), *Biochem. J.*, **117**, 405–407.

[7] Morse, D.E. and Horecker, B.L. (1968), *Adv. Enzymol.*, **31**, 125–181.

[8] Hoek, J.B., Ernster, L., de Haan, E.J. and Tager, J.M. (1974), *Biochim. biophys. Acta* **333**, 546–559.

[9] Storer, A.C. and Cornish–Bowden, A. (1974), *Biochem. J.*, **141**, 205–209.

[10] Commission on Biochemical Nomenclature (1973), *1972 Recommendations on Enzyme Nomenclature*, Elsevier, Amsterdam.

[11] Brown, A.J. (1902), *J. Chem. Soc.*, **81**, 373–388.

[12] Michaelis, L. and Menten, M.L. (1913), *Biochem. Zeitschr.*, **49**, 333–369.

[13] Woolf, B., quoted by Haldane, J.B.S. and Stern, K.G. (1932), *Allgemeine Chemie der Enzyme*, Steinkopff Verlag, Dresden, p. 119.

[14] Lineweaver, H. and Burk, D. (1934), *J. Amer. Chem. Soc.*, **56**, 658–666.

[15] Wilkinson, G.N. (1961), *Biochem. J.*, **80**, 324–332.

[16] Dowd, J.E. and Riggs, D.S. (1965), *J. biol. Chem.* **240**, 863–869.

[17] Cleland, W.W. (1967), *Adv. Enzymol.*, **29**, 1–32.

[18] Eisenthal, R. and Cornish–Bowden, A. (1974), *Biochem. J.*, **139**, 715–720.

[19] Markus, M., Hess, B., Ottaway, J.H. and Cornish–Bowden, A. (1976), *FEBS Letters*, **63**, 225–230.

[20] de Miguel Merino, F. (1974), *Biochem. J.*, **143**, 93–95.

[21] Briggs, G.E. and Haldane, J.B.S. (1925), *Biochem. J.*, **19**, 338–339.

[22] van Slyke, D.D. and Cullen, G.E. (1914), *J. biol. Chem.*, **19**, 141–180.

[23] Cornish-Bowden, A. (1976), Principles of Enzyme Kinetics, pp. 142–152, Butterworths, London.

[24] Schwert, G.W. (1969), *J. Biol. Chem.*, **244**, 1278–1284 and 1285–1290.

[25] Bates, D.J. and Frieden, C. (1973), *J. biol. Chem.*, **248**, 7878–7884 and 7885–7890.

[26] Heinrich, R. and Rapoport, T.A. (1974), *Eur. J. Biochem.*, **42**, 89–95.

[27] Rapoport, T.A., Heinrich, R., Jacobasch, G. and Rapoport, S. (1974), *Eur. J. Biochem.*, **42**, 107–120.

[28] Harris, J.I., Meriwether, B.P. and Park, J.H. (1963), *Nature, Lond.*, **198**, 154–157.

[29] Piszkiewicz, D., Landon, M. and Smith, E.L. (1970), *J. Biol. Chem.*, **245**, 2622–2626.

[30] Krebs, H.A. and Eggleston, L.V. (1940), *Biochem. J.*, **34**, 442–459.

[31] Wassarman, P.M. and Major, J.P. (1969), *Biochemistry*, **8**, 1076–1082.

[32] Dixon, M. (1953), *Biochem. J.*, **55**, 170–171.

[33] Gundlach, H.G., Stein, W.H. and Moore, S. (1959), *J. biol. Chem.*, **234**, 1754–1760.

[34] Crestfield, A.M., Stein, W.H. and Moore, S. (1963), *J. biol. Chem.*, **238**, 2413–2420 and 2421–2428.

[35] Herries, D.G., Mathias, A.P. and Rabin, B.R. (1962), *Biochem. J.*, **85**, 127–134.
[36] Dixon, M. (1953), *Biochem. J.*, **55**, 161–170.
[37] Michaelis, L. and Davidsohn, H. (1911), *Biochem. Zeitschr.*, **35**, 386–412.
[38] Alberty, R.A. and Massey, V. (1954), *Biochim. biophys. Acta,* **13**, 347–353.
[39] Dixon, M. and Webb, E.C. (1964), *Enzymes* 2nd edn. pp. 118–128, Longmans, London.
[40] Hakala, M.T., Glaid, A.J. and Schwert, G.W. (1956), *J. biol. Chem.*, **221**, 191–209.
[41] Adams, M.J., Buehner, M., Chandrasekhar, K., Ford, G.C., Hackert, M.L., Liljas, A., Rossmann, M.G., Smiley, I.E., Allison, W.S., Everse, J., Kaplan, N.O. and Taylor, S.S. (1973), *Proc. Natn. Acad. Sci.*, **70**, 1968–1972.
[42] Dalziel, K. (1957), *Acta chem. scand.*, **11**, 1706–1723.
[43] Alberty, R.A. (1953), *J. Am. chem. Soc.*, **75**, 1928–1932.
[44] Engel, P.C. and Dalziel, K. (1970), *Biochem. J.*, **118**, 409–419.
[45] Theorell, H. and Chance, B. (1951) *Acta chem. scand.*, **5**, 1127–1144.
[46] Dalziel, K. (1963), *J. biol. Chem.*, **238**, 2850–2858.
[47] Dalziel, K. and Dickinson, F. M. (1966), *Biochem. J.*, **100**, 34–46.
[48] Morrison, J.F. and James, E. (1965), *Biochem. J.*, **97**, 37–52.
[49] Massey, V., Gibson, Q.H. and Veeger, C. (1960), *Biochem. J.*, **77**, 341–351.
[50] Massey, V. and Veeger, C. (1963), *Ann. Rev. Biochem.*, **32**, 579–638.
[51] Visser, J., Voetberg, H. and Veeger, C. (1970), In *Pyridine Nucleotide Dependent Dehydrogenases.* (ed.) H. Sund, pp. 359–373, Springer Verlag, Berlin.
[52] Koster, J.F. and Veeger, C. (1968), *Biochim. biophys. Acta,* **151**, 11–19.
[53] Dixon, M. and Kleppe, K. (1965), *Biochim. biophys. Acta,* **96**, 368–382.
[54] Doudoroff, M., Barker, H.A. and Hassid, W.Z. (1947), *J. biol. Chem.*, **168**, 725–732.
[55] Frieden, C. (1959), *J. biol. Chem.*, **234**, 2891–2896.
[55a] Silverstein, E. and Sulebele, G. (1973), *Biochemistry*, **12**, 2164–2172.
[56] Keleti, T. (1965), *Acta physiol. Hung.*, **28**, 19–29.
[57] Macfarlane, N. and Ainsworth, S. (1972), *Biochem. J.*, **129**, 1035–1047.
[58] Fromm, H.J. (1967), *Biochim. biophys. Acta,* **139**, 221–230.
[59] Dalziel, K. (1969), *Biochem. J.*, **114**, 547–556.
[60] Fromm, H.J. (1975), *Initial Rate Enzyme Kinetics,* Springer-Verlag, Berlin.
[61] Fromm, H.J. (1963), *J. biol. Chem.*, **238**, 2938–2944.
[62] Cleland, W.W. (1963), *Biochim. biophys. Acta,* **67**, 188–196.
[63] Boyer, P.D. (1959), *Archs. Biochem. Biophys.*, **82**, 387–410.
[64] Boyer, P.D. and Silverstein, E. (1963), *Acta chem. scand.*, **17**, S195–S202.
[65] Silverstein, E. and Sulebele, G. (1969), *Biochemistry*, **8**, 2543–2550.
[66] King, E.L. and Altman, C. (1956), *J. phys. Chem.*, **60**, 1375–1378.
[67] Ainsworth, S. (1975), *J. theor. Biol.*, **50**, 129–151.
[68] Volkenstein, M.V. and Goldstein, B.N. (1966), *Biochim. Biophys. Acta,* **115**, 471–477.
[69] Hurst, R.O. (1969), *Can. J. Biochem.*, **47**, 941–944.
[70] Ainsworth, S. and Kinderlerer, J. (1976), *Int. J. Bio-medical Computing,* 7, 1–20.
[71] Denton, R. and Pogson, C.I. (1976), *Metabolic Regulation,* Chapman and Hall, London.
[72] Cohen, P. (1976), *Control of Enzyme Activity*, Chapman and Hall, London.
[73] Monod, J., Changeux, J.-P. and Jacob, F. (1963), *J. molec. Biol.*, **6**, 306–329.
[74] Koshland, D.E. Jr., Némethy, G. and Filmer, D. (1966), *Biochemistry,* **5**, 365–385.
[75] Hill, A.V. (1913), *Biochem. J.*, **7**, 471–480.
[76] Adair, G.S. (1925), *J. biol. Chem.*, **63**, 529–545.

[77] Pauling, L. (1935), *Proc. natn. Acad. Sci.*, **21**, 186–191.
[78] Perutz, M.F., Rossmann, M. G., Cullis, A. F., Muirhead, H., Will, G. and North, A.C.T. (1960), *Nature*, **185**, 416–422.
[79] Novick, A. and Szilard, L. (1954), In *Dynamics of Growth Processes*, (ed.) E. J. Buell, p. 21, Princeton U. Press, Princeton.
[80] Umbarger, H.E. (1956), *Science*, **123**, 848.
[81] Yates, R.A. and Pardee, A.B. (1956), *J. biol. Chem.*, **221**, 757–770.
[82] Umbarger, H.E. and Brown, B. (1958), *J. biol. Chem.*, **233**, 415–420.
[83] Helmreich, E. and Cori, C.F. (1964), *Proc. natn. Acad. Sci.*, **51**, 131–138.
[84] Gerhart, J.C. and Pardee, A.B. (1962), *J. biol. Chem.*, **237**, 891–896.
[85] Changeux, J.-P. (1962), *J. molec. Biol.*, **4**, 220–225.
[86] Koshland, D.E. Jr. (1958), *Proc. natn. Acad. Sci.*, **44**, 98–104.
[87] Monod, J., Wyman, J. and Changeux, J.-P. (1965), *J. molec. Biol.*, **12**, 88–118.
[88] Dalziel, K. (1958), *Trans. Faraday Soc.*, **54**, 1247–1253.
[89] Ingraham, L.L. and Makower, B. (1954), *J. Phys. Chem.*, **58**, 266–270.
[90] Ferdinand, W. (1966), *Biochem. J.*, **98**, 278–283.
[91] Sweeny, J.R. and Fisher, J.R. (1968), *Biochemistry*, **7**, 561–565.
[92] Rabin, B.R. (1967), *Biochem. J.*, **102**, 22C–23C.
[93] Wyman, J. (1948), Adv. Protein Chem., **4**, 407–531.
[94] Gerhart, J.C. and Schachman, H.K. (1965), *Biochemistry.*, **4**, 1054–1062.
[95] Patte, J.-C., Truffa-Bachi, P. and Cohen, G.N. (1966), *Biochim. biophys. Acta*, **128**, 426–439.
[96] Crawford, I.P. (1975), *Bact. Revs.*, **39**, 87–120.
[97] Lornitzo, F.A., Qureshi, A.A. and Porter, J.W. (1975), *J. Biol. Chem.*, **250**, 4520–4529.
[98] Koberstein, R. and Sund, H. (1973), *Eur. J. Biochem.*, **36**, 545–552.
[99] Panagou, D., Orr, M.D., Dunstone, J.R. and Blakley, R. L.(1972), *Biochemistry*, **11**, 2378–2388.
[100] Levitzki, A. and Steer, M.L. (1974), *Eur. J. Biochem.*, **41**, 171–180.
[101] Wyman, J. (1972), *Curr. Top. Cell. Regul.*, **6**, 209–226.
[102] Engel, P.C. and Dalziel, K. (1969), *Biochem. J.*, **115**, 621–631.
[103] de Vijlder, J.J.M. and Slater, E.C. (1968). *Biochim. biophys. Acta*, **167**, 23–34.
[104] Conway, A. and Koshland, D.E. Jr. (1968), *Biochemistry*, **7**, 4011–4023.
[105] Levitzki, A. and Koshland, D.E. Jr. (1969), *Proc. natn. Acad. Sci.*, **62**, 1121–1128.
[106] Teipel, J. and Koshland, D.E. Jr. (1969), *Biochemistry*, **8**, 4656–4663.
[107] Engel, P.C. and Ferdinand, W. (1973), *Biochem. J.*, **131**, 97–105.
[108] Kirschner, K., Eigen, M., Bittman, R. and Voigt, B. (1966), *Proc. natn. Acad. Sci.*, **56**, 1661–1667.
[109] Hoggett, J.G. and Kellett, G.L. (1977), *Biochem. Soc. Trans.*, in press.
[110] Ogawa, S. and McConnell, H. (1967), *Proc. natn. Acad. Sci.*, **58**, 19–26.
[111] Changeux, J.-P. and Rubin, M.M. (1968), *Biochemistry*, **7**, 553–560.
[112] McClintock, D.K. and Markus, G. (1968, 69), *J. biol. Chem.*, **243**, 2855–2862 and **244**, 36–42.
[113] Scarano, E., Geraci, G. and Rossi, M. (1967), *Biochemistry*, **6**, 192–201.

Index